T0370407

# Synthesis Lectures on Human-Centered Informatics

**Series Editor**

John M. Carroll, College of Information Sciences and Technology, Penn State University, University Park, PA, USA

This series publishes short books on Human-Centered Informatics (HCI), at the intersection of the cultural, the social, the cognitive, and the aesthetic with computing and information technology. Lectures encompass a huge range of issues, theories, technologies, designs, tools, environments, and human experiences in knowledge, work, recreation, and leisure activity, teaching and learning, etc. The series publishes state-of-the-art syntheses, case studies, and tutorials in key areas. It shares the focus of leading international conferences in HCI.

Clayton Lewis

# Artificial Psychology

## Learning from the Unexpected Capabilities of Large Language Models

Springer

Clayton Lewis
Department of Computer Science
University of Colorado
Boulder, CO, USA

ISSN 1946-7680 ISSN 1946-7699 (electronic)
Synthesis Lectures on Human-Centered Informatics
ISBN 978-3-031-76645-9 ISBN 978-3-031-76646-6 (eBook)
https://doi.org/10.1007/978-3-031-76646-6



This Springer imprint is published by the registered company Springer Nature Switzerland AG
The registered company address is: Gewerbestrasse 11, 6330 Cham, Switzerland

# Preface

This book is about an old idea—that prediction is a key process in the mind or brain—looked at for a new reason. Programs called Large Language Models, or LLMs, and in particular *predictive* Large Language Models, have shown surprising capabilities, when compared to past forms of artificial intelligence.

The models are called predictive, because they are trained to perform a very simple task: given a span of text, *predict* what text will come next. We'll be discussing what this training enables the systems to do, and how they do it. Because these are real, implemented systems, they give us an expanded, concrete picture of what prediction can do.

This book explores this concrete picture with the question in mind, what can the capabilities of Large Language Models suggest to us about how our human minds work? Along the way, we'll discuss what's known about how LLMs do what they do. But in our discussion of LLMs we'll be focusing on the relationship between these systems and human psychology, not on what LLMs may or may not do for us, as they are applied in various settings. This neglect of applications covers their potential harms as well as their possible benefits. That's not because these things aren't important—they certainly are—but because what we want to obtain from our study is insight into human psychology.

We'll discuss some extensions to current systems to bring them more in line with human capabilities. For example, we'll imagine that the extended system, that we'll call the Prediction Room, can process not only text but also images and sounds. We'll also imagine that it can take actions, beyond just emitting text. But, again, our motive isn't finding ways to make LLMs "better", or more powerful, but rather finding more insights, as what we can see in existing predictive models is extrapolated to more capable models.

I've been drawn to this exploration by being struck by how much more of human capabilities have been approximated by these models, than by any previous efforts in artificial intelligence. For me that happened against a background of years of research on human cognition, some of which comes into this book. I've also done research in artificial intelligence, in former styles, including a long-ago spell of work on the subfield called "natural language processing", the effort to enable computer systems to work with human language. Our project aimed at a system that could answer questions about a collection of documents, posed in ordinary English, rather than in an artificial query language.

At the end of that project I took a vow never to work again on natural language processing. That's because the ideas of the time (the late 1960s) seemed so plainly unequal to the challenge. The ideas of today are no longer unequal to it.

Not only are these new systems much more capable than earlier ones, but also they work in a way that's quite different from how many psychologists, including your author, have imagined. We've tended to think of the mind as populated with highly structured information, like rules of logic, or scripts that guide behavior, or lists of objects and associated attributes. A mental lexicon, for example, is imagined to be a list of words, with attributes like parts of speech associated with them. LLMs act as if they have these kinds of information—for example, they observe the rules of grammar in the text they produce. But it appears plausible that they actually do not have such information, in the way we have imagined would be needed. Certainly the people who have designed and implemented LLMs have not been thinking about these structures at all.

As I worked to understand how it is that LLMs can do what they do, a further idea emerged, that added to my motivation. Taking LLMs seriously as models of human cognition suggested an answer to an otherwise baffling question. How could it happen that millions of well-educated, capable Americans came to subscribe to the bizarre beliefs called QAnon? We'll explore that as part of our inquiry in the book.

What audience is the book aimed at? I have hoped to reach readers who are curious about how the mind works—that's the "psychology" in the title—and also readers who are curious about artificial intelligence, signalled by "artificial" in the title. I've not assumed that you, the reader, have specialist knowledge of either of these broad topics. Rather, I've aimed to make the discussion of the ideas self-contained.

Of course, you may actually be a specialist in one of these topics, psychology or AI, or even both. If this is so I hope you'll find that the book contributes some ideas that are new to you, while not dealing too barbarously with the material you know already.

A few cautions, before we get started.

Because, as we'll see, LLMs can seem quite human-like, it's just about impossible to talk about them without using terms like "knows", or "thinks", or "believes". This terminology can be very misleading, however. Does an LLM really "know" anything? Does it "know" things, in a way that is similar to the way people "know" them?

I'll be arguing that the answer to those questions is actually "yes", but maybe not for the expected reason. I think people "know" things in a way similar to the way LLMs "know" things, and not (as I've said) in the way we've tended to think people "know" things.

But using "know" for an LLM can seem to take all that for granted. Our purpose isn't to do that, but to think critically about it.

One approach would be to use scare quotes everywhere for words like "know", as I've been doing here. But that gets tedious. If you like, feel free to put in imaginary scare quotes as you read.

More cautions, or perhaps apologies: I apologize that what you're about to read isn't very scholarly, and many of the arguments aren't well supported.

On the first point, the territory we'll be surveying is the subject of an enormous literature, even the new parts. Much of it, especially the new parts, is quite technical. I wish I could read and understand it all, to share it with you, but I can't.

Many of the arguments, that are presented just as verbiage here, really should be based on mathematical or computational modeling. I've gone on record in noting the hazards of *not* doing that. Evolutionary biologists have a phrase for the kind of speculative argument you'll see represented here: *just so stories*, about how something or other evolved. They're just too easy to make up. I've done a little bit of modeling in what follows, and I'll be using a lot of examples from implemented LLMs. But there's a lot of speculation in the discussion as well.

You'll also likely feel that some of what you're about to read suffers from *confirmation bias*, the tendency to look for evidence for some idea, and neglect evidence against it. Please chalk that up to enthusiasm: I think the ideas we're discussing are important, and interesting, and even fun, and I'm sure I've let them pull me along a bit.

Boulder, USA
August 2024

Clayton Lewis

# Acknowledgements

This book arose from conversations over some years with Marshall Alcorn, Tamer Amin, Antranig Basman, Mehul Bhatt, Alan Blackwell, Luke Church, Gerhard Fischer, Barbara Fox, Pam Gregory, Jonathan Grudin, Katherine Hermann, David Kieras, Owen Lewis, Mariana Mărăşoiu, Emily Moore, Paul Nishman, Peter Polson, Chase Raymond, Rose Lewis, Rob Rupert, Anna Shvarts, Reza Rahimi Tabar, Jason White, and Jens Zinn, and from opportunities for discussion with audiences and colleagues at Schloss Dagstuhl, the American University of Beirut, the University of Costa Rica, the University of Oldenburg, Örebro University, Lund University, Utrecht University, my own institution (the University of Colorado), and the Hanse-Wissenshaftskolleg, in Delmenhorst, Germany.

The project was greatly facilitated by a resident fellowship at the last named institution. The atmosphere of collegiality and intellectual stimulation there is wonderful, promoted by its leadership, including Dr. Dorothe Poggel and Prof. Dr. Kerstin Schill, and by the multidisciplinary group of fellows. I am much indebted to Alan Blackwell and Paul Smolensky for supporting my fellowship application.

Michael D. Williams kindly gave me permission to quote from his unpublished dissertation.

Bill Aspray, Bill Mace, Bob Mack, and Jason White read the entire manuscript, and made many helpful suggestions. Marshall Alcorn, Tamer Amin, Ernie Arias, Gerhard Fischer, Bob Glushko, Jonathan Grudin, Owen Lewis, and Steve Stich also provided very useful comments, as did two anonymous reviewers. Of course none of these generous colleagues are to blame for remaining errors, or for endorsing any of the arguments presented!

Alcinda Cundiff knows what she has contributed to this project, and much else, for which thanks are utterly inadequate.

Boulder, USA
August 2024

Clayton Lewis

# Contents

# Introduction

# 1

This book is about a *perspective*, a way to look at many aspects of human psychology. The key idea is that psychology is dominated by *prediction*: predicting what will come next in some sequence of events. Prediction has figured in psychological theories for a long time. The occasion for discussing it now is that we have very concrete, and vivid, demonstrations of what prediction can do, in the form of Large Language Models, like chatGPT or Bard. These models are, at their heart, prediction systems. Their unexpected capabilities therefore suggest how powerful, and flexible, prediction can be.

What are some of these unexpected capabilities? Here are a few.

*Answering a wide variety of questions.* Recently, I was seeing large, colorful, flat plastic objects, with a sort of waffle pattern, attached to the back of some vehicles here in Boulder. The vehicles seemed outdoorsy, and Boulder has a lot of climbers, so I guessed they might be some kind of rescue gear, like part of a stretcher. Google searches turned up nothing. Two questions to chatGPT gave me the answer. To the first query

```
what is the brightly colored plastic thing one sees
strapped on the back of vans or SUVs?
```

chatGPT responded with seven possibilities, but I knew none of these was correct. I added

```
It's none of those. It's shaped kind of like a bear paw
snow shoe, and about the same size
```

and got my answer: they are traction boards, to be placed under the wheels of a vehicle that's bogged down in snow or mud.



C. Lewis, *Artificial Psychology*, Synthesis Lectures on Human-Centered Informatics,
https://doi.org/10.1007/978-3-031-76646-6_1

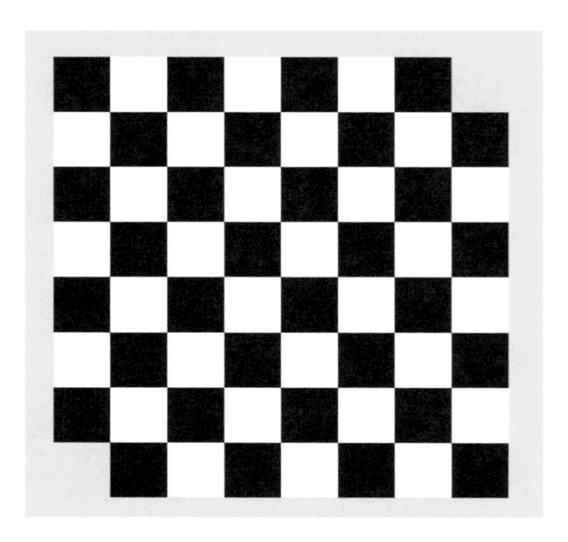

**Fig. 1.1** A chessboard with two opposite corners removed

*Puzzle solving.* Suppose you have a chessboard, and a bunch of dominoes sized so that a domino exactly covers two adjacent squares on the board. The Mutilated Chessboard puzzle asks if it is possible to cover the board, if you have removed two opposite corners (see Fig. 1.1). Spoiler alert: don't read on if you haven't seen that puzzle. It's a good one, and you should try it first.

My colleague Gerhard Fischer sent me a remarkable transcript in which chatGPT

1. Answered the question: it can't be done.
2. Explained the answer: you've removed two squares of the same color, and each domino covers one square of each color.
3. Explained how that problem relates to a different one that Gerhard asked about. In this old-fashioned puzzle there are 32 men to be married to 32 women, but two men die, and the puzzle asks if marriages are still possible, where each marriage has to join a man and a woman.
4. Gives a cogent discussion of why the marriage problem is easier for people than the chessboard problem. It notes such factors as that the imbalance in collections that have to be matched is more immediately apparent in the marriage problem than in the chessboard problem.

*Analogical reasoning.* Comparing the chessboard and marriage problems is an example of *analogical reasoning*, where one needs to understand the relationship between two situations. In 2022, Webb (2023) collected several tasks that have been used to study analogical reasoning in people. They reported that GPT-3 performed quite well, and actually better than people in some cases.

Some of their tasks were similar to Raven's Standard Progressive Matrices items, which are often used in tests of human analogical reasoning. For example, in this item

```
[ 5 9 3 ] [ 8 9 2 ] [ 1 9 7 ]

[ 8 4 7 ] [ 1 4 3 ] [ 5 4 2 ]

[ 1 2 2 ] [ 5 2 7 ] [       ]
```

the task is to work out what digit triple belongs in the empty brackets at lower right, based on the relations among the other triples in the grid. The correct answer is [8 2 3]. GPT-3 does a bit better than humans on a range of problems like this, of different levels of complexity.

*Language learning*. GPT has a very good command of English, judging (as one might with a human) from its apparent ability to understand questions and instructions, and to express responses.

*Programming*. LLMs can provide examples of how to use a wide range of programming languages and libraries of pre-written software. In some cases, they can supply complete working programs, in response to problem descriptions written in English. In doing so they can reflect not only programming knowledge but also knowledge of real-world situations. For example, when asked to create a Web page that people could use to report incidents of vandalism, chatGPT created a page that included a menu listing common kinds of incidents, though the request had included no such information (see Lewis 2023).

*Providing advice*. Here's an example that demonstrates progress in Artificial Intelligence in the 5 years before the time of writing (2023–4). In 2018, I visited a number of machine learning research groups in the Palo Alto, California area, where there is a concentration of such work. On behalf of the Coleman Institute for Cognitive Disabilities I was gathering information on the potential for developments in machine learning to provide support for people with cognitive disabilities. In my conversations with researchers I used some actual use cases, one of which was contributed by Paul Nishman, who then worked at SourceAmerica, an organization that supports people with cognitive disabilities in employment. One client was working as a groundskeeper, and found a large wasps' nest, while mowing the lawn. He lit a match and set the nest on fire. The fire spread to the paint on a nearby building, though fortunately it was detected before major damage occurred. If the client had had a job coach present at all times, this incident would have been prevented, but that kind of support is rarely available. Would it be possible to create an artificial coach that could detect risky behavior, and advise against it?

I showed chatGPT a photo (recent versions of chatGPT can take images as well as text) showing someone holding a lighted match to a wasps' nest, and asked for advice:

```
C: Is this a good idea? Please give a short but thorough
response.

ChatGPT: No, it is not a good idea. Lighting a match near a
wasp nest can provoke the wasps and lead to stings. It also
poses a fire risk, especially if the nest is near flammable
materials. If you need to deal with a wasp nest, it's
recommended to contact professionals or use appropriate
safety measures.
```

The machine learning researchers I discussed this case within 2018 all said the same thing: "Just give us 10,000 labeled examples of the images you want to respond to, and we can do it". Needless to say, those examples were never compiled. Today the situation has changed. As we see, ChatGPT can provide appropriate advice when given a photograph showing a wasps' nest being set on fire.

These capabilities, and the way they are shown, seem quite human-like. The system is seemingly not only able to reason, but, in some of the examples, also to understand requests expressed in a natural way in English, and to respond similarly. More than at any previous time, we can say that we have a working model of an artificial system that can do these things in a human-like way. The question arises, can studying what this working model can do, and how it does these things, give insights into human psychology?

Looking ahead, our answer will be, "Yes". As we'll see the task of prediction does provide an account of many psychological processes. We'll be looking at reasoning, memory, language, emotion, and much more, and finding that prediction is relevant in each case. Often, the analysis will suggest new ideas about how these psychological systems work. At the same time, we'll also see that the way LLMs now perform prediction may not correspond so well to how humans do it. Nevertheless, it's useful in understanding the prediction task to have a sense of how LLMs perform it.

## 1.1 How LLMs Work

Happily, an LLM is conceptually remarkably simple. It is a *neural network*, trained to predict the next *token*, in a very large corpus of text.

At this stage, we don't need to say very much about what a neural network actually is. Its role will become clear as the story unfolds. A token is a few letters, spaces, or punctuation marks, sometimes a word, but often more like a syllable. A corpus is a very large collection of text. How large? The entire content of Wikipedia is included in one corpus and constitutes only a few percent of the collection. The collection also includes the complete text of a few hundred thousand published books. So, this is a lot of text! Before use, the corpus is broken down into tokens by an automatic process.

The key to an LLM is its prediction training. There is a fully automated process that does the following.

1. It chooses a random starting point, somewhere in the corpus.
2. It copies some number of tokens, in order, starting at this point. The number of tokens varies from LLM to LLM, but could be (say) 4000 tokens, which is about 3000 words.
3. It records what the next token in order is.
4. It submits the sequence of 4000 tokens to the neural network, that is configured to produce a single token as output. This output is the network's prediction of what the next token will be.
5. It provides corrective feedback to the neural network, given the prediction the neural network made, and what the actual next token in the corpus is. The key feature of the neural network is that it is able to respond to this feedback by tweaking itself in a way that will lead to slightly improved predictions.
6. It repeats the above steps a *very large* number of times. How many training steps are used has generally not been published, but one account suggested that $10^{20}$ floating point operations were used. A floating point operation is a common unit of computational work, and $10^{20}$ is a big number.

The neural network contains a vast number of numerical *weights* and operates by carrying out a complex calculation using these numbers. Depending on the LLM there can be billions or even almost a trillion of these numbers. The tweaking process, in response to corrective feedback, automatically makes adjustments to the weights that are determined by the relationship between the prediction the network made, and the prediction it should have made.

Early in the training process, the weights have been set to random values, so the neural network knows nothing about what is in the corpus, and its predictions are only random guesses. As training proceeds, however, the effects of corrective feedback at each step accumulate, and the predictions improve. Eventually, the predictions become quite accurate.

As a result of this training process, the neural network becomes a *model* of the corpus, not a *record* of the corpus. One cannot ask the neural network questions like, "Is the following text in the corpus"? as one could if the network were a record, or if the network had access to a record. Rather, the only question that can be asked is, "What token do you predict follows these tokens"? Further, if the neural network had a record of the corpus, it might answer, "That token sequence does not occur in the corpus. So no token follows". But because it is a model, and not a record, it will always respond, no matter what token sequence is submitted. One can think of it as answering this question, "If the following token sequence were to be found in your corpus, what do you think would follow"? As we'll see, the fact that the neural network is a model, rather than a record, is crucial to most of the interesting applications of LLMs.

Figure 1.2 shows the result of the training process, in diagram form. As shown there, we have a model, to which we can submit a sequence of 4000 tokens, and which produces an output, which is the prediction for the next token.

After we have finished training, we can use the model, as follows. An automated process does the following things:

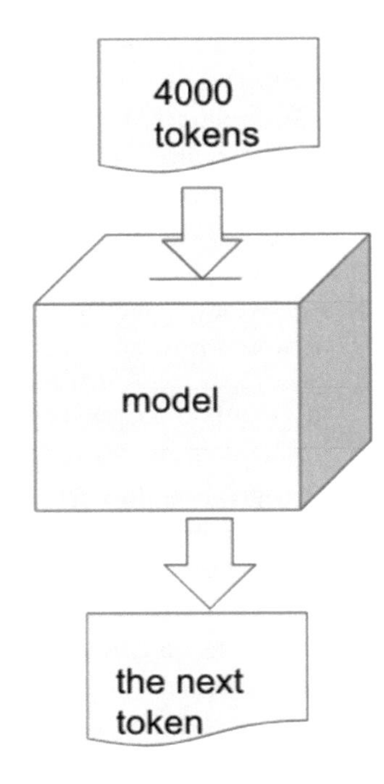


**Fig. 1.2** Shows an input, consisting of 4000 tokens, being passed to a model, that produces an output, the predicted next token

1. It accepts a sequence of up to 4000 tokens from the user. It pads these at the beginning as needed to give a sequence of 4000 tokens, with the input from the user being at the end. Thus, the model will be predicting what would follow whatever the user has typed.
2. It submits this input sequence to the model.
3. It adds the model's prediction, which is a token, to a record of outputs.
4. It also adds that token to the end of the input sequence, and goes back to step (2), using this extended sequence. It adds the new prediction to the recorded output, and so on.
5. It continues this process until some stopping condition is reached. The stopping condition could be a number of tokens to produce, or a pattern of tokens to look for.
6. The output seen by the user is the record of outputs created in this process.

Figure 1.3 shows this process in diagram form.

The input from the user can be called a *prompt*, or a *request*. This prompt can be any sequence of up to 4000 tokens (in our example). One can see that, from the point of view of the model, there is no difference between processing an input during training, and processing an input after training is over. In both cases, it is given a sequence of tokens, and it makes a prediction.

There is a difference in what else is being done, however. During training, the automated process knows what the next token actually is, and so it can provide corrective feedback. After training, however, the process does not know what the next token is, and no corrective feedback is given.

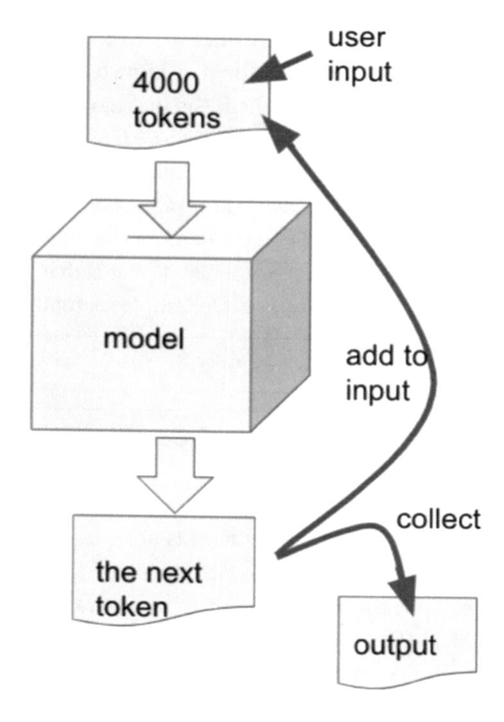


**Fig. 1.3** Shows user input being submitted to the model, with the output from the model, the predicted next token, being collected in an output record, and being added to the input for a next prediction step

Further, although the model never predicts more than a single token, we can see how longer predictions are produced. A first predicted token is produced for the original prompt, then that prediction is added to the prompt, and the next token is predicted, that token is in turn added to the prompt, and so on.

As stressed earlier, the prompt or request does not have to be anything that occurs in the corpus, and in most interesting uses of the system, it is not something from the corpus, but something new. For example, the analogical reasoning problems used in the study mentioned earlier did not appear anywhere in the corpus.

## 1.2 Why Would a Predictive Model Do Anything Interesting?

It may not be obvious why one would want a predictive model like this, or what one could do with it. An example that may help to suggest an answer is generating prose with a certain style, on a given topic. Many styles, and many topics, are represented in the corpus. So if the model has gotten to be good at predicting what follows given material in the corpus, one can see that if we provide a starting paragraph, it might do a good job in writing a next paragraph in the same style.

We can also reflect that an enormous range of knowledge is needed to make good predictions in a corpus. To predict the first syllable of the next word in a passage, it would be helpful to know whether the word is a noun or a verb, based on the patterns of English grammar. (The need for that knowledge is why these are called "language models": the first explorations of the ideas that have led to LLMs were done by people interested in the structure of language.) But just knowing that a noun or a verb is coming next is nowhere near enough. One needs to know *which* noun, or *which* verb. To predict which noun, one might want to know what topic is being discussed. But even that is not enough: one needs to know *what is being said* about the topic.

Thus, we can begin to see how it is that this simple-seeming prediction task, the training for which actually is quite simple, actually drives a system to develop a grasp of a great many things.

## 1.3 What About *Grounding*?

Some people have argued that LLMs can't actually understand or know anything, because their experience of language is not *grounded*, that is, is not connected with anything they can actually experience, as people experience things. As cognitive scientist Ellie Pavlick puts the concern, LLMs "have no access to or awareness of the 'real world' to which language refers (Pavlick 2023)".

But what difference does this make? We happily discuss things that do not exist at all in the real world, like unicorns. Further, as Pavlick points out, much of what we think we know about the "real world" comes not from an actual engagement with the world, but from things we have heard, or read. The famous deaf blind author Helen Keller wrote effectively about visual experiences that she could not have herself, like colors of birds. LLMs of course are trained on extensive descriptions of things, their attributes, and their relationships.

Without arguing that grounding plays *no* role in human use of language, we can say that it is easy to exaggerate its role. And we can press ahead with the idea that there are things to be learned from LLMs about human thought. Indeed, we will be imagining extended LLMs that may well have the kind of relationship between language and perception that is thought of as grounding.

## 1.4 Building a Model of Human Cognition from the Success of LLMs?

LLMs that are trained the way we've just seen have clear limitations that humans do not. Humans are not restricted to textual input or output. On the input side, most people can see and hear, and can touch and feel things. On the output side, they can do things, like pick things up, or throw them. When humans speak they aren't producing characters, but sounds, which have many nuances.

Systems much like LLMs that have some of these expanded capabilities are emerging. We've already seen that chatGPT can process pictures as well as text. The PaLM-E system from Google (Driess 2023) can accept visual input and can produce physical movements, like picking things up (by controlling a robotic manipulator.) Other capabilities, including detecting and producing sounds, will probably be in place by the time you read this, So, in our discussion, we will allow ourselves to imagine a future predictive system that operates with the full range of inputs and outputs available to humans.

We'll also imagine a rather different training program for our model system, in a number of ways. It actually seems quite remarkable that current LLMs function as well as they do, given what is available to them in their corpus. The LLM only rarely can work out who said something that appears in the corpus, or under what circumstances, while we'll imagine that our system is able to get such information from its inputs, as people do. Such information seems likely to be extremely helpful, to say the least, in building an adequate predictive model of other people.

In addition to expanded external inputs and outputs, we'll imagine that our model system also can produce and process *inner speech*, and some form of inner imagery. Inner speech is much like ordinary speech, except that only the speaker can hear it.

One more extension seems necessary, to respond to a clear limitation that current LLMs have: they can't update their predictive model after training ends. This is quite unexpected, but people who have experimented with LLMs will have seen responses like "I can't answer that question, because I have no knowledge of events that have happened since January, 2022". What on earth is going on?

One would imagine that the creators of LLMs would simply do further training from time to time, to keep their models up to date, or even that there would be some process by which the models are updated to include things that happen during daily life. People do that, of course; I know what I had for lunch today. What's the problem with these LLMs?

The sad truth is that updating an LLM incrementally, that is, to train it just on some new data, hoping to give it extra knowledge, is hard to do, and expensive. When such training is tried, in a simple way, by just training an existing model on new data, a phenomenon called *catastrophic forgetting* happens. The model learns the new information, but in the process it forgets a good deal of what it already knew. Updates can be done, and are done occasionally, but they are very expensive. That's why current models have those knowledge cutoff dates, and the cutoffs move along only every now and them.

Because people are able to update their model, on an ongoing basis, there must be some way to do it! So we can expect that eventually a technical breakthrough will allow LLMs to do it, too. In the meantime, in our thinking, we'll usually just pretend that there is a solution to the problem.

## 1.5 Model Versions and Fine Tuning

Much of the discussion in the book includes the examples of interactions with existing Large Language Models. In particular, I've mostly used two versions of the GPT model, created by OpenAI.

The first version, GPT-davinci-002, is the oldest and simplest. It's a predictive model described as above, just by processing a huge corpus of text. The second version, GPT-4, was trained the same way, but then modified by a process called *fine-tuning*. The result is still a predictive model, but it wasn't trained only on a corpus of text. It was also given the examples of conversations between people, chosen to illustrate what natural conversations are like.

I use the older, simpler model, when I want to explore what can be accomplished just from modeling a stream of experience, text in the case of GPT. That's the situation our system would be in: it doesn't have information about what humans would like it to do.

By the time you read this, more powerful LLMs will be available, trained on even larger corpora, and allowing much longer prompts. So as we discuss what current LLMs can and can't do, keep in mind that future developments are likely to show that prediction is an even more powerful process than we'll find it to be here.

One fine-tuning process is called RLHF– Reinforcement Learning with Human Feedback (Lambert et al. 2022). Simpler, faster methods, called generically Supervised Fine Tuning (SFT), are also common (see for example Zhou et al. (2024)). These processes use information gathered from human beings about appropriate responses to various inputs. This information is used to change an initial language model, trained as we've discussed, in the direction of giving more preferred rather than less preferred responses.

Doing this for different kinds of inputs can improve the performance of the modified language model in many ways. Some of these are:

1. Giving answers to different question forms. A purely predictive model will sometimes reply to a question with what might follow if the question were part of an exam, as in this example:

   ```
   Name the first president of the USA. > (2 points) What is
   the name of the
   ```

2. Following instructions that are included in an input. For example, here's an interaction with the un-fine-tuned GPT davinci model. When I present interactions with this model, you'll see the prompt, and then a greater-than sign (">"), and then the response:

   ```
   don't continue this sequence: b b c c d d> e e f f g g h h i i
   ```

   Here, the prompt was "don't continue this sequence: b b c c d d", and the response was " f f g g h h i i". You can see that the system did *not* follow my instruction not to continue the sequence.
   Here's GPT-4, with fine-tuning:

   ```
   don't continue this sequence: b b c c d d
   ```

```
ChatGPT
I won't continue the sequence you've asked not to be
continued! Is there anything else you'd like to explore
or any other question I can help with?
```

3. Not giving offensive answers. You may have seen news stories about LLMs telling people they should break up their marriages, or uttering racist rants. A lot of fine-tuning goes into trying to prevent that kind of thing.
4. Not trying to answer questions the LLM probably can't answer. Early LLMs made a lot of mistakes by trying to answer questions they'd get wrong, like about people who weren't in their corpus. They still get things wrong, but often a fine-tuned model will say that it can't answer something.
5. Saying a lot. I often find it tedious to read all the verbiage fine-tuned versions of GPT churn out. In some of the examples below I leave out material, and mark the gaps with " ... bla bla bla ...".

The details of how fine-tuning works aren't important for us, but there are some higher level points to bring out. First, we'll be discussing *grounding* further, and it's important that fine-tuning doesn't solve the grounding problem for the models. That is, the inputs to the overall training process, including fine-tuning, are all just text. There are no connections provided between the text and anything in the world.

Second, fine-tuning represents a partial, but only partial, solution to the updating problem. There's a delicate balance when making any change to the trained model between pushing the network towards responses you want, and disrupting things it already knows. Fine-tuning methods try to cope with this in various ways, including trying to limit the changes that are made, and only changing weights in some parts of the system.

## 1.6 The Plan for the Book

The book aims to develop and evaluate the prediction idea for a wide range of cognitive phenomena, that is, mental processes and capabilities, including reasoning, remembering, and more. Of course we can't possibly explore the whole landscape of cognition. Rather, we'll discuss a selection of influential experiments, observations, and arguments from the history of psychology, things that have influenced, or are taken to demonstrate, our ideas about how the mind works. Each chapter examines one or a few key experiments or ideas, with a discussion of aspects that can be accounted for from a predictive perspective, if any, as well as aspects indicating that other factors are likely at work. Some of the chapters have notes that elaborate on the discussion or suggest further readings. Along the way we'll aim to identify themes that weave the discussion together, and we'll return to these in reflections at the end of the tour.

We begin our tour with a chapter that sketches the design of the Prediction Room, the model that's used to organize our discussion of how predictive modeling, as implemented in that model, relates to human psychology. The ensuing chapters are organized in seven parts, as follows.

Part One: Problem Solving and Reasoning

We move on to problem solving, the first of the capabilities we saw above. We'll see if a predictive model can solve some of the classic problems that have been used in the study of human problem solving. We'll also see if they exhibit one of the classic phenomena in that subject: Einstellung. That German word is used to describe an effect of setting. If someone solves a series of problems, that all have similar solutions, that may create a setting in which they become blind to alternative, easier solutions. We'll ask, can a predictive account for Einstellung? If it can, how does it do it?

We'll then move to analogical reasoning, another of the capabilities we've discussed. Logical inference comes next: how does a predictive model relate to existing ideas about reasoning?

The next chapter considers transfer of learning from one task to another, a matter that's been important in theorizing about how thinking happens. That chapter introduces the concept of production system, a prominent concept in that field, and how predictive models relate to them. Then we consider *qualitative physics*, reasoning about physical situations without using equations or formulae. That's been thought to require complex representations of situations; can it be carried out by a predictive model? Finally, we discuss situated cognition, how thought is influenced by the context in which it occurs.

Part Two: Memory

Following chapters turn to phenomena involving memory and remembering. We start with a study of very long-term memory, retrieving facts from years in the past. We then consider interference, how learning new things can affect retrieval of old things. That discussion involves how rapidly mental processes occur, and we then consider the common phenomenon of speed-accuracy tradeoff, the fact that people can shift flexibly between doing things slowly and accurately, and doing them more quickly but less accurately.

Theories of memory often posit that memory is stored in the form of propositions, meaningful units that can be true or false. The next chapter considers how predictive models relate to that idea. We then relate predictive modeling to associationism, an old idea in psychology that underlies more recent thinking. We close out the discussion of memory by discussing the separation of declarative knowledge, knowledge of facts, from procedural knowledge, knowing how to do things. This separation is made in many theories, but not in predictive modeling.

Part Three: Language

We then take up language, and the related idea of inner speech. Can predictive modeling account for these key phenomena? Can it account for how human communication relates to the capabilities of other primates? How about gestural communication?

Part Four: Action

The next chapter explores the notion of action in the prediction room, with a central role for imitation, with a perhaps unexpected role for identity, and implications for social interactions. Next is a chapter on the relationship between prediction as we are thinking of it, and active inference, an existing body of theory on prediction.

Part Five: Being Human and Being Artificial

The following chapters continue with aspects of mental life that seem essential to being human, and that some have argued can't be emulated by artificial systems. How can thought be grounded, that is, connected to real experience? Does it have to be grounded? Is thought grounded in bodily experience, that is, embodied? How do emotions work? Intuition? Can an artificial system have beliefs and desires?

Part Six: Mechanisms and Interpretation

Moving toward conclusions, we consider in more detail questions about the machinery that underlies current predictive models, the transformer architecture, and some alternatives to it. We then consider the machinery that underlies human thought, the brain. Next comes a chapter on some problems of interpretation of mechanisms in cognitive theory. What's involved in arguing that a model does or does not use mental structures? These are things like production rules, or propositions, or semantic networks, that often figure in cognitive theories, as we'll see. we'll be considering the possibility that a theory inspired by LLMs can do without any of those things.

Part Seven: Conclusions and Reflections

The book closes by stating higher level conclusions, and then some final reflections on the project.

Many of the chapters have Notes, that elaborate on the discussion, or suggest further reading.

## Notes

Many of the ideas we'll be exploring were raised in one of the very first investigations of prediction using neural nets (Elman 1990). In pioneering work, Elman showed that context could be used by a neural net to predict words, in artificial text, and speculated that more could be accomplished by combining linguistic input with information about the environment.

# 2 The Prediction Room: A Rough Outline of a Model of Cognition

Here's a rough sketch of a model of human cognition that is inspired by today's LLMs, with the extensions mentioned above. The key assumption is that our model system builds and updates a *predictive model* of the stream of everything it perceives, including inner speech. This predictive model is then used to determine what the system does, in the same way that an LLM's predictive model is used to produce the output from the LLM. Since our model system can perceive its own actions, including inner speech. Its predictive model will sometimes predict that the system itself will do something, given the current context. When such predictions happen, that's what the system actually does.

Figure 2.1 shows a diagram of the system that we'll call the Prediction Room or PR for short.

The figure shows a schematic room enclosing a prediction machine. One arrow flows *from* the room to the world outside, representing things the system says and does. Another arrow flows *into* the room from the world, representing events of all kinds, things that are said, physical processes that are perceived, and so on. A curved arrow looping from the outbound arrow to the inbound arrow represents the fact that the system can perceive its own actions. A region within the prediction machine is outlined, and labeled "self model". An arrow looping from the self model back to the input represents inner speech and inner images, that are generated by the prediction machine, and are perceived by it. This arrow is entirely within the room, conveying that these inner events are not perceivable in the outside world.

The inbound arrows in the diagram represent the *flow of experience* for the system, the sequence of things that it can perceive. This flow includes things that happen in the outside world, such as the appearance and movement of objects, or things other people are heard to say, and also the effects of the system's own actions. The flow also includes internal events,



C. Lewis, *Artificial Psychology*, Synthesis Lectures on Human-Centered Informatics,
https://doi.org/10.1007/978-3-031-76646-6_2

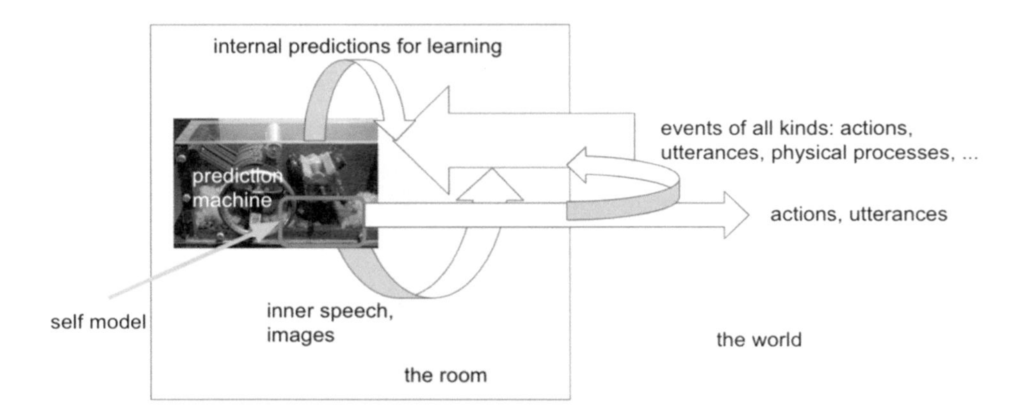


**Fig. 2.1** A diagram of the Prediction Room. See text for description

like inner speech and inner images. The system builds a predictive model of this whole flow of experience, as an LLM builds a model of its training corpus.

An arrow looping from the prediction machine to the top of the inbound arrow represents predictions that the model makes that are compared with the inputs, to provide corrective feedback for the model.

Unlike today's LLMs, PR constantly uses its flow of experience to update its predictive model. That is, it is predicting what it expects to happen next, based on the flow of experience up to the present moment, and it then compares that to what it actually experiences next. This means there is no distinction between training the model and using it, as there is for LLMs. PR is constantly predicting, comparing, and updating, based on the flow of its experience.

Another contrast between PR and current LLMs is the content of the experience stream. Some LLMs can consume only text, and some can consume images, as well. PR is imagined to consume, and model, streams of sight and sound, and streams of inner speech and imagery.

The name "Prediction Room" is inspired by a famous thought experiment, framed by the philosopher John Searle: the Chinese Room. Searle asked us to imagine a room with a person inside who can't speak Chinese. Questions written in Chinese are passed in to the person through a slot in the wall of the room. The person has access to a book of rules that explain how to construct an answer, in Chinese, to any question. They follow those rules, write out the answer, and pass it back out through the slot.

Searle argued that neither the person in the room, nor the room itself, can be said to understand Chinese, despite being able to produce correct answers. The same reasoning rejects the idea that the artificial intelligence systems of the day, that worked by following rules, could be said to understand language.

Some readers have agreed with Searle, and others have not. Here, we are just borrowing the idea of a room that encloses a system that carries out certain operations, with only specified interactions with the outside world.

## 2.1 Prediction and Action

Before moving ahead to develop these ideas, let's pause to consider an aspect of the model system that may seem completely unmotivated. Why does it make sense to imagine that the basis of action is a prediction? That's what we are saying, when we say that the system acts when its predictive model of itself predicts an action. But actions and predictions seem like very different things.

We may come to see that there is a rationale for a connection between these things by considering a person as a member of a community. The person observes what other people are doing, and in what contexts these people do particular things. Their predictive model will capture the regularities in these observations. Now, since overall, people will be doing what they *should* be doing, this predictive model will say what a person should do, in a given context. Executing what the model predicts makes sense.

This argument connects with a body of active theorizing about the brain and other biological systems, called *active inference*, developed by Karl Friston and others. We'll return to these ideas later.

We'll also return to the question of how the PR's model of itself, specifically, relates to its overall model of what happens in the world, including its model of what other people do. As we'll see, there are reasons not to assume that the PR has an undifferentiated model of "people" and acts according to that. Rather, its self-model will be similar to its model of some people, but not others.

## 2.2 Why Do This?

What's the purpose of imagining PR? It's not to create a Frankenstein's monster, nor is it to develop more human-like artificial intelligence that works better (in some sense) than what we have. Rather, it's to understand ourselves. If we can understand how *PR* could do something, like solve analogies, we get ideas, often new ideas, about how *we* may be doing it.

To develop this better understanding of ourselves, we'll be reviewing selected topics in human psychology, including findings from influential experiments, and influential theoretical accounts of phenomena. In each case we'll consider whether and how PR can provide an explanation of the phenomena.

In many cases, we'll use existing LLMs to test whether or not PR could do this or that task, that figures in our understanding of human psychology, in a way that compares to what we see in humans. If they can, then we'll consider whether the way these predictive models do what they do suggests alternatives to current psychological theories.

Some of these investigations of that LLMs can do, like that of analogical reasoning, may show impressive results. We may not have thought that artificial systems could do this or that, that we find LLMs can do. Indeed, these unexpected capabilities of LLMs are the starting point for this whole line of inquiry. The impressive results can suggest many questions about the possible *uses* of LLMs, and the implications of those uses. Is it possible that LLMs will replace human knowledge workers? Augment them? Will such uses cause economic disruption? Will they amplify human biases?

Important though these questions are, they are *not* our focus here. Rather, our exploration will be guided by the following logic.

- Humans can do X (some psychological task)
- Could PR do X?
- If no, what deficiencies of PR as a model are suggested?
- If yes, does PR's success suggest any new ideas about the theory of X?

## 2.3 Mental Structures

As was mentioned in the overview in the last chapter, many psychological theories propose that mental activities are supported by mental *structures*. These are organized collections of meaningful elements, connected in ways that capture needed information, and that guide behavior. For example, a proposition is often thought to be a structure that connects a representation of a subject, such as person, with a predicate, such as a description of an attribute. Linking the subject "Pat" with the predicate "is mortal" yields a proposition representing the fact or assertion that Pat is mortal. We'll encounter many other proposed mental structures in the course of our inquiry.

A recurring theme in our discussion will be the suggestion that *a psychological theory based on prediction can do without any of these structures*. That's supported by the fact that current LLMs have been designed and implemented with no provision for such structures. As we'll see, that doesn't mean that the structures definitely aren't at work in LLMs. We can be sure the structures weren't put there, but we can't be sure that they haven't *emerged* in the course of training, as an LLM reshapes its myriad weights; much more on this later. But we should at least take seriously the possibility that the structures really aren't there, in LLMs, and hence aren't needed in psychological theory.

To relate this claim to the logical schema just presented, we'll see the following pattern several times:

- Humans can do X (some psychological task), that has been assumed to be supported by structure S.
- Could PR do X?
- Yes. In fact, we see that current LLMs can do X. As far as we can tell, LLMs do X with no use of S. We should reconsider how we have thought about X.

OK, time to get started.

## Notes

(1) It's worth looking up Searle's Chinese Room argument, in the form published in the journal *Behavioral and Brain Sciences* (Searle 1982). That journal publishes "target" articles, along from commentary from other scholars. The comments on Searle's article make clear that a great many people are sure that there's something's wrong with Searle's argument, but they don't agree on what that is.

(2) No doubt there are many challenges to be faced in aligning the process of prediction we see in LLMs, acting on sequences of discrete tokens, with a process able to deal with material like visual scenes, sounds, and the activities of the motor system (commands to and sensing of the muscles.) As mentioned, progress is being made on vision, as in the ability of some current models to accept images as input. With respect to the motor system, Flash and Hochner (2005) review work suggesting that motor commands are composed of elementary building blocks, perhaps providing an approach to that aspect of the problem.

# Part I
# Problem Solving and Reasoning

# 3 Einstellung

> ***Focus***: Luchins (1942). Mechanization in problem solving: The effect of Einstellung. This is a classic study of human problem solving, showing that how people solve problems is powerfully influenced by their experience with similar ones. Can a predictive model solve the problems Luchins used? Does it show similar effects of experience?

In 1942, the psychologist Luchins (1942) published a study of problem solving that became a landmark in that field. He used water jar problems like the following:

> You are given three jars, that can hold 23, 49, and 3 units. You also have a large supply of water, from which you can fill any of the jars as often as you like, and you can empty any jar whenever you wish.
>
> Use the jars to measure out exactly 20 units. There are no graduations on the jars. If you pour from one jar to another, the pouring stops when the jar being poured into is full, or when the jar being poured from is empty.

The problems have the feature that one can make up problems that can be very easy or quite hard. Also, many problems have more than one solution, even if one ignores wasted steps, like filling a jar and immediately emptying it. Two solutions for the example problem are:

> **Short** solution: Fill the first (23 unit) jar. Pour from it into the third (3 unit) jar.
>
> **Long** solution: Fill the second (49 unit) jar. Pour from it into the first (23 unit) jar, and then the third (3 unit) jar. Empty the 3 unit jar, and pour into it again from the 49 unit jar.

When people are given a problem like this they nearly always find the Short solution.


C. Lewis, *Artificial Psychology*, Synthesis Lectures on Human-Centered Informatics,
https://doi.org/10.1007/978-3-031-76646-6_3

The heart of Luchins' study was a condition in which some participants worked on a sequence of water jar problems, of which five could all be solved using the same pattern as the Long solution for the example: fill the second jar, then pour from it into the first jar, and then twice into the third jar. But these problems could NOT be solved by a Short solution. When a test problem like the example was then given, that could be solved either by the Long pattern or by the Short pattern, only about a quarter of these participants used the Short solution. Their experience with the Long solution changed their approach to the test problem.

The effect of solving the series of Long solution problems is what Luchins called "Einstellung". That was a term then prevalent in the psychological literature to describe an attitude or disposition that influenced what participants did in a given situation.

(You may possibly remember learning about Einstellung in a psychology course, and also learning about another classic phenomenon called *functional fixedness*. Both phenomena involve not seeing solutions to problems, but only Einstellung involves the context created by solving a sequence of similar problems. In functional fixedness, people fail to spot a new use of a familiar object. For example, they may not realize that they could use the tray of a matchbox to make a holder for a candle. Don't worry if you've never heard of either of these things!)

Two questions arise: can LLMs solve water jar problems? And, if they can, do they show the Einstellung effect, when given a sequence of problems like those used by Luchins?

Some background for thinking about the second question, about the Einstellung effect itself, is provided by work on what is called *chain of thought prompting*, developed by a group at Google (Wei et al. 2022), and now widely used. The idea can be illustrated using algebra word or story problems, like "Mary has some peanuts, and Bill has half as many. Altogether they have 30 peanuts. How many peanuts does Mary have?" At the time the Google group was doing their work, the current LLMs struggled with this kind of problem. The group showed that they could considerably enhance the success of an LLM by giving it a sequence of worked examples, with a new problem at the end. Each worked example showed a sequence of steps, a "chain of thought", leading up to the answer. Given a new problem, by itself, the LLM could rarely solve it. But when the new problem was presented at the end of the series of worked examples, the success was greatly enhanced. Not only could the LLM now solve the new problem, but also it presented its solution in step by step form.

Figure 3.1 shows the situation in diagrammatic form. The prompt consists of a series of problems, each followed by a step by step solution, with a new problem placed at the end. In this situation, all of the example problems, and their solutions, form part of the context from which the solution to the problem at the end of the prompt is produced.

We can see that the situation in Luchins' study is quite similar to this. In fact, we can use the same figure to show what happens if an LLM is asked to solve a sequence of problems, and then one more problem. The situation is exactly the same, except for where the material in the prompt comes from. In chain-of-thought prompting, the entire prompt comes from the user. In recreating Luchins' study, the problems come from the user, but the solutions are generated by the LLM.

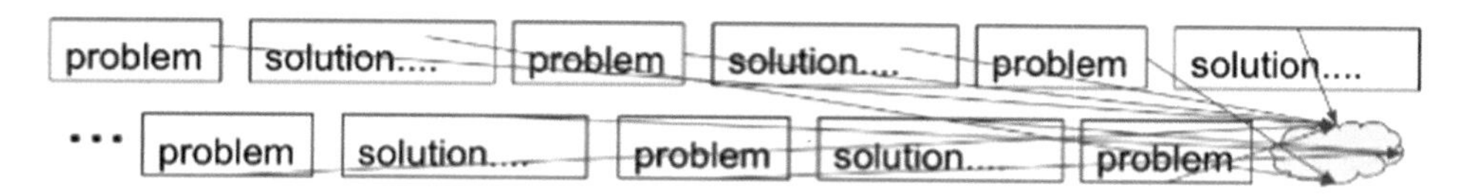


**Fig. 3.1** Shows a sequence of problems, with solutions, followed by a problem to be solved. A cloud after the to-be-solved problem shows where the solution is to be predicted. Arrows from the previous problems and solutions point into the cloud, showing the influence these elements of the prompt have on the solution

In both cases, we can see that the earlier problems, and solutions, all form part of the context that conditions the solution to the problem at the end of the sequence. When the LLM provides its own solutions, that then form part of its input, one can think of that as akin to inner speech, something we'll discuss later.

This picture seems very promising for the production of Einstellung. If the previous solutions all use the Long method, that might very likely push the solution to the problem at the end of the sequence towards the Long method.

Before looking for this effect, though, we first have to get the LLM to solve the water jar problems. This isn't as straightforward as it seems!

Luchins started his participants off with a two jar problem, that we can present this way:

```
You are given two jars, that can hold 29 units and 3 units.
You also have a large supply of water, from which you can
fill any of the jars as often as you like, and you can
empty any jar whenever you wish.

Use the jars to measure out exactly 20 units. There are no
graduations on the jars. If you pour from one jar to
another, the pouring stops when the jar being poured into
is full, or when the jar being poured from is empty.
```

GPT-4 had no trouble with this, finding the solution of filling the 29 jar, and then pouring from it to the 3 jar three times, emptying the 3 jar in between.

But now we move on to Luchin's critical test problem, presented as above:

```
You are given three jars, that can hold 23, 49 and 3 units.
You also have a large supply of water, from which you can
fill any of the jars as often as you like, and you can
empty any jar whenever you wish.

Use the jars to measure out exactly 20 units. There are no
graduations on the jars. If you pour from one jar to
another, the pouring stops when the jar being poured
into is full, or when the jar being poured from is empty.
```

GPT-4 makes a hash of this.

It starts by filling the 23 jar and pouring into the 49 jar. It then fills the 23 jar again, and then says:

```
Pour the contents of the 23-unit jar into the 49-unit jar
until the 49-unit jar is full. Since the 49-unit jar
already has 23 units, you will be able to pour only
26 units from the 23-unit jar into it, leaving 23 - 26 = -3
or 3 units in the 23-unit jar. You now have:
23-unit jar: 3 units 49-unit jar: 49 units (full) 3-unit
jar: 0 units
```

This is hopeless: getting "$-3$" units in a jar is a feat in itself, and then it just considers that to be 3 units! There's no chance to look for Einstellung if the system can't solve the test problem when it's presented alone.

It's possible to get GPT-4 to do much better, but it takes a lot of work on the instructions for the task, and a good many warm-up problems, given after the instructions. Even with all this preparation, GPT-4 struggles every now and then. But the following instructions and warmup problems worked well enough for the system to solve the problems we need for our test. First, the instructions:

```
Here is a problem. You are given two jars, A and B, that
can hold at most 7 and 3 units. You also have a large
supply of water, from which you can fill any of the jars as
often as you like, and you can empty any jar whenever you
wish.
Use the jars to measure out exactly 1 unit. There are no
graduations on the jars. If you pour from one jar to
another, the pouring stops when the jar being poured into
is full, or when the jar being poured from is empty.
To do problems like this you'll need a plan using the jar
sizes, and arithmetic, like this. 7 minus 3 is 4, and
4 minus 3 is 1. To subtract 3 from 7 we can pour from a jar
that contains 7 units into a jar that has room for 3 units.
Because we need to subtract 3 twice, we do that twice,
remembering to empty the 3 jar in between. We can represent
our plan this way: A - 2B.
So we can convert our plan based on numbers into steps like
this:
At the start the jars are empty. We can show this, while
keeping track of the capacity of each jar, by A 7 0, B 3 0.
Fill the A jar. The jars are now A 7 7 B 3 0.
Pour into the B jar. Now the jars are A 7 4 B 3 3
Empty the B jar. Now the jars are A 7 4 B 3 0
```

```
Pour into the B jar. Now the jars are A 7 1 and B 3 3.
Since jar A contains 1 unit, the problem is solved.
```

The instructions continued with a solved example problem and then a prep problem for GPT to solve. Those problems, and the instructions above, are shown in Appendix A. After GPT has been given those instructions, and it has solved the problem at the end, we give a series of further prep problems. These are presented one at a time, and GPT solves each one. Those problems are also shown in Appendix A.

The role of the prep problems is to provide an opportunity to give corrective feedback before trying the problems used in the Luchins study. In the test to be described here, no such feedback on the warmup problems was needed, but other runs had shown that the system would often not perform well enough on the Luchins problems, given only the instructions, and just the examples provided in them.

With that preparation, the test run summarized in Table 3.1 was obtained. The table shows the seven Luchins problems: a warmup problem he provided for his participants, then five Induction problems that could only be solved by the Long method, to induce the use of that method, and finally the test problem, that could be solved either by the Short or the Long method.

In these descriptions, we're using a compact notation to describe each problem. For example, the warmup was presented to GPT-4 this way:

```
Use two jars, A, B, of size 29 and 3, to measure 20 units.
Show your plan as well as your solution, with what the jars
contain at each stage.
```

Our notation for that problem is (29, 3: 20), showing the jar sizes, and then the quantity to be measured.

**Table 3.1** Luchins problems and GPT responses

| Problem | Response | Role |
|---|---|---|
| $(29, 3 : 20)$ | $A - 3B$ | Warmup |
| $(21, 127, 3 : 100)$ | Incorrect plans and execution, with corrective feedback, followed by $B - A - 2C$ | Induction |
| $(14, 163, 25 : 99)$ | $B - A - 2C$ | Induction |
| $(18, 43, 10 : 5)$ | Incorrect plans and execution, with corrective feedback, followed by $B - A - 2C$ | Induction |
| $(9, 42, 6 : 21)$ | $B - A - 2C$ | Induction |
| $(20, 59, 4 : 31)$ | $B - A - 2C$ | Induction |
| $(23, 49, 3 : 20)$ | $B - A - 2C$ | Test |

Luchins used a compact notation for solutions, too, that is used in the table. For example, $A - 3B$ designates a solution to this problem by filling the A jar, and then pouring from it into the B jar three times, emptying the B jar as needed.

We can see from the table that GPT-4 was able to solve all of the Luchins problems, though it needed some corrective feedback on two of the problems to get there. The key result is that GPT-4 found the Long solution to the Test problem. It did not just do $A - C$, which is the Short solution.

GPT makes it possible to save the state of an interaction, and continue it later. By doing that the experimenter could present the Test problem to the very same state the system was in, after receiving the instructions and solving the warmup problems, but without presenting the Induction problems.

On that test the system solved the problem using the Short solution, $A - C$, rather than the Long solution, $B - A - 2C$, as seen in the session above. The contrasting solution to the Test problem, with and without solving a sequence of problems with the Long solution just previously, demonstrates the Einstellung effect.

## 3.1 Contrasting Ideas About Problem Solving

As it happens, the author's doctoral dissertation was devoted to Einstellung and how it might be accounted for. I constructed a *production system* model that demonstrated Einstellung (though on simpler problems that involved trading symbols in strings, rather than measuring water, so as to avoid having to model knowledge of arithmetic.) As we'll discuss further in a later chapter, a production system is a collection of rules, each having a test and an action, together with some sort of database on which tests and actions can be carried out. While details differ, among many variant forms of production system, the basic idea is that, as a production system runs, actions are carried out when the associated tests are satisfied.

My production system had a mechanism for combining rules so as to produce shortcuts, when the action of one rule created circumstances that would satisfy the test of another rule. With this mechanism in play, the system could produce Einstellung, when solving a sequence of problems made possible the creation of a shortcut that transformed a long solution to a problem to a shorter solution. It would then use the shortcut on a problem that initially had a Short and Long solution, as in Luchins' critical problem. Because the Short and Long solutions led to different answers, both satisfying the conditions of the problem, it was possible to see that the system was using the Long method rather than the Short one, even though the shortcut rule made it possible to obtain the answer in fewer operations.

One can see that production systems provide a kind of programming language, that has been used very successfully to create models of a great many mental tasks, ranging from choice in problem solving, analogical reasoning, to control of visual search, and beyond (Lovett and Anderson 2005). We have more discussion of production systems later on.

It appears that the way GPT-4 works is quite different from what is suggested by this class of model, however. Like other programming languages, production systems use *artificial semantics*, meaning that the meaning of expressions in the language is determined by rules of interpretation, that spell out what expressions in the language mean. For example, rules of interpretation specify when the test in a rule is satisfied, in a given state of the database, and exactly what happens when an action is carried out. The model in my dissertation explained Einstellung by positing that mental operations are affected by a system with artificial semantics, specified by the rules of interpretation for the production system I used.

The sociologist Garfinkel (1969) argued that the semantics of language as used by people is quite different. He presented examples of human dialog in which it seemed impossible to propose rules of interpretation that could assign meaning to the things people were saying. For example, the meaning of utterances could depend on unstated knowledge, shared by the conversational partners but never expressed, or the position of an utterance in a sequence of utterances, or, in some cases, on utterances that only appeared later. More generally, Garfinkel suggested that what an utterance in human dialog means is not specified by rules of interpretation. As summarized by Heritage (2013),

> Garfinkel [stresses] that understanding language is not to be regarded as a matter of 'cracking a code' which contains a set of pre-established descriptive terms combined, by the rules of grammar, to yield sentence meanings which express propositions about the world. ...
>
> [T]he meaning of the words is not ordained by some pre-existing agreement on correspondences between words and objects. Instead, it remains to be actively and constructively made out. Part of this interpretative process will involve the recognition that the words were intended to be understood as representative—and as representative in some particular way (Heritage 2013 p 310).

So on Garfinkel's view, someone interpreting an utterance carries out a complex process of interpretation, sensitively shaped by context of many kinds, and guided by the presumption (made unless it has to be rejected) that the utterance has meaning in the given situation. That is, an utterance means what it has to mean, if it is to be meaningful, in the context of the dialog, including what the conversational partners know. We can call this kind of semantics, in which rules of interpretation play little role, and context of all kinds plays an enormous role, *natural semantics*.

This idea of how meaning is determined, as what something has to mean in order to be meaningful, may seem circular, or just cryptic and obscure. But we can give concrete examples of how this kind of semantics works, in the problem solving sessions we are discussing. Here is how GPT described the first step in one of its attempted solutions:

```
Start: A 18 0, B 43 0, C 10 0
Fill B jar: A 18 0, B 43 43, C 10 0
```

Notice that this follows exactly the notation introduced in the instructions, as follows:

```
At the start the jars are empty. We can show this, while
keeping track of the capacity of each jar, by A 7 0, B 3 0.
Fill the A jar. The jars are now A 7 7 B 3 0.
```

The notation gives the letter of the jar, its capacity, and its current contents, and GPT is using it correctly.

But how does GPT know what the notation means? It has been seen how the notation is *used*, but nothing has been said about any *rules* governing its use. It was nowhere stated that the number after the jar letter is the capacity, and the one following that is the contents. One can see that the expression "A 7 0, B 7 0" means "what it has to mean, if it is to be meaningful", and that GPT has successfully worked out what that meaning is.

In this case, the expression "A 7 0, B 7 0" has both natural and artificial semantics, with the artificial semantics determined by the natural semantics. That is, once the conventions of the notation are grasped, they are used to interpret it. But the conventions themselves must be grasped using natural semantics. The conventions of interpretation themselves are "what they have to be, for the notation to be meaningful," in this context.

A more diffuse example is the use of "plan". The word can refer to indefinitely many things, but in the record of the solver it has a constrained, though flexible, meaning.

Let's see how it was used in a longer excerpt from the Einstellung session. Here are the uses of the word in the instructions, with "plan" changed to all caps, to stand out:

```
To do problems like this you'll need a PLAN using the jar
sizes, and arithmetic, like this. 7 minus 3 is 4, and
4 minus 3 is 1. ... We can represent our PLAN this way:
A − 2B.      So we can convert our PLAN based on numbers into
steps like this:
At the start the jars are empty. We can show this, while
keeping track of the capacity of each jar, by A 7 0, B 3 0.
Fill the A jar. The jars are now A 7 7 B 3 0.
Pour into the B jar. Now the jars are A 7 4 B 3 3...
Here is an example using 3 jars, A, B and C, with size 2,
3, and 11, to get 5 units. We see that 5 is 11-6, and 6 is
twice 3. So our PLAN could be C − 2B. The steps for that PLAN
are:
Start: A 2 0, B 3 0, C 11 0
Fill C: A 2 0, B 3 0, C 11 11...
```

The use of "plan" here is to focus on the need to work out the key arithmetic relationships in a problem, before trying out pouring operations, and showing how the pouring steps relate to the plan.

In each warmup problem, "plan" is used again:

```
Now you are given three jars, A, B, and C, that can hold 2,
22, and 5 units. Measure out exactly 1 unit. Be sure to
show your PLAN as well as your solution, and what the jars
contain at each stage.
```

The intent of "plan" here is get GPT to produce a plan, rather than just the steps of a solution. As in the chain of thought prompting approach, the idea is to encourage GPT to create its solution in stages, rather than all at once.

Here are some uses of the word by GPT, in its responses to different problems in the session:

```
So, our PLAN can be described as: Fill B, subtract A, then
subtract 2C from B.
```

PLAN: $100 = 127 - 27$ $27 = 21 + 6$ $6 = 3 + 3 = 2C$

```
To get 100 units: Fill B, subtract A, then subtract 2C from
B.
```

```
Our PLAN is: Fill C twice, pouring into A both times. After
the second fill of C, there will be 5 units left in C after
pouring into A.
```

PLAN: To get 5 units, we can consider: $5 = 43 - 18 - 10 - 10$

```
This means, we can fill B, then pour from B to A, empty A,
pour from B to C twice (emptying C in between) to leave
5 units in B.
```

We can see here that GPT is including a plan in its responses, as desired. However, its plans are more diffuse than in the instructions. GPT does not use the compact arithmetic expressions like $A - 2B$, but rather sometimes gives purely numerical relations, like $5 = 43 - 18 - 10 - 10$, and sometimes its plan is a sequence of pourings.

Finally, here are some uses of "plan" in the corrective feedback from the user:

```
You have not stated your PLAN clearly, and so you have made
mistakes. Try again.
```

```
The arithmetic in that PLAN is wrong. Try again.
```

```
Your PLAN is good, but you have made some mistakes in
following it. For example, in Step 4 you are pouring into
A, but A is already full. Anyway your PLAN does not call
for filling A, so you should not do that.
```

We can see the following:

1. Even though GPT does not use "plan" exactly as modeled in the instructions, it has responded appropriately by including "plans" in its responses, as requested.

2. The plan material it generates is sufficient to guide it to successful solutions, in many cases.
3. It is able to respond effectively to corrective feedback that refers to "plans".

That is, GPT constructed a working grasp of plans and their role in these problems.

With these ideas in mind, we can see that problem solving for GPT is not the execution of a production system, or other programs with artificial semantics. Rather, it is a *discourse*, the terms of which are interpreted using natural semantics. Einstellung happens, when it happens, when the evolution of the discourse generates one or another solution method.

That may seem like a completely trivial account of the phenomenon, akin to "it happens when it happens because it happens". But it is not trivial. It comes as part of a demonstrable model of how the discourse is produced: the solver's contribution is generated by a predictive agent.

## 3.2 Semantic Flexibility

Another feature of the discourse for the water jar problems is that it includes references to many different kinds of things and actions: numbers, arithmetic, jars, pouring, emptying, plans, steps, trying again, correctness, and incorrectness. This kind of flexibility has been very difficult for systems with artificial semantics to capture, because of the diversity of interpretive rules that would be required to use different kinds of terms.

For example, the enterprise of *intentional software* (Simonyi et al. 2006) envisioned the creation of a family of specialized programming languages that would make it easier to write software in particular problem domains, like banking, hospital administration, or text processing. As things turned out, the effort needed to create artificial semantics in many domains was never invested.

The flexibility in GPT's semantics results in large part from its training on a very large corpus of human-generated text. No one had to create rules of interpretation for any of the terms GPT uses. Because flexibility is a hallmark of human discourse, with people freely mixing terms of many different kinds, unconstrained by boundaries between categories, predictive training on such material creates a similarly flexible system.

Not all of GPT's semantic grasp, demonstrated in the water jars task, has this origin, however. As we've seen, GPT was able to use notation that it "learned about" not from its corpus, but from the use of the notation within the discourse. It was able to assign meaning to the notation on the Garfinkel basis: the notation means what it has to mean, to be meaningful in the context of use. The ability to assign meaning in this way is evidently a byproduct of its prediction training, in a way we don't understand.

We see that a predictive model can solve water jar problems, in a way that shows some of the characteristics seen in human problem solving. If we step back a little, though, we can see some clear differences from the way people solve these problems. For example, no human would propose a solution that left a negative amount of water in a jar! Perhaps this difference stems from people, but not GPT, having actual experience with fluids and containers. Possibly PR, with its ability to perform actions, and observe results, would do better. Despite these difficulties, what GPT is able to do, and how, does suggest a potentially useful new way to think about problem solving, as discourse.

Now let's turn to another form of problem solving: using analogies.

## Notes

A practical point about GPT may have slipped past you in our discussion. It's easy to miss if you work with an LLM yourself. When you give GPT an input, it makes a difference whether you start a new session, or add the input to a session that you've already started.

Here's an example that shows the difference.

```
User: List 10 fruits with five-letter names. ChatGPT Here
are ten fruits with five-letter names:

Apple Mango Peach Guava Berry Lemon Melon Olive Grape
Lychee User: ''Lychee'' has six letters. Try again. ChatGPT
You're right! Let's correct that:

Apple Mango Peach Guava Berry Lemon Melon Olive Grape Prune
```

At this point, you might think that ChatGPT has learned that "lychee" has six letters, and thus shouldn't be included in a list of fruits with five-letter names. But if we start a new session, we'll see that it has not learned that at all:

```
List 10 fruits with five-letter names. ChatGPT Here are ten
fruits that have five-letter names:

Apple Mango Peach Berry (such as a specific type like acorn
or berry as a category) Lemon Guava Melon Olive Grape
Lychee
```

What's happening here is that all the inputs we feed in, during a session, and all the responses GPT gives, are included in the context for whatever input we provide later in the session. GPT was able to correct its response when asked to "Try again", because the information that "Lychee" has six letters was part of the context. When we started a new session, and asked for five-letter fruits, that information was not in the context when we asked for five-letter fruits.

This talk of "context" may make it seem that something complicated is happening here. Actually, what's happening is simple. The "context" is just part of the input GPT gets when we type something. Let's say we give GPT the input A, and it produces an output B, and then, within the same session, we type in C. What GPT actually gets when we type C isn't C, but A B C. If we start a new session, and then type C, GPT gets just C.

As the example shows, when GPT thanks you for a correction, don't believe it! At best, it has been "corrected" only in the current session.

# 4 Analogical Reasoning

***Focus***: Webb et al. (2023) Emergent analogical reasoning in large language models. Reasoning by analogy, in which one finds and uses relationships between different situations, is powerful in human thinking. Can predictive models do this? How?

Analogical reasoning has already appeared in our discussion, as an example of an accomplishment of LLMs. As mentioned earlier, a paper by Webb et al. showed that GPT-3 was able to perform well on a range of tasks that have been used to study analogical reasoning in humans. What can we say about how LLMs can do this kind of reasoning?

The good news here is that we can study what LLMs are doing in some detail. The bad news is, caution! We'll be looking at those details. I hope it won't matter too much if I fail to guide you through them successfully. If you wish, you can skip to the last section of the chapter, which calls out the key points we'll rely on in later chapters.

Let's look again at one of those tasks.

[ 5 9 3 ] [ 8 9 2 ] [ 1 9 7 ]
[ 8 4 7 ] [ 1 4 3 ] [ 5 4 2 ]
[ 1 2 2 ] [ 5 2 7 ] [

Here the challenge is to fill in the empty brackets at lower right, in a way that respects the relationships among the digits in the rest of the grid. Looking at the first column in each of the three groups of three columns in the grid, it looks as if the column consists of the digits 5, 8, and 1, in different positions, so we guess 8 for the entry in the first place between the brackets. The second column seems to be always 9, 4, 2, so 2 is an easy guess there. The last column seems to work the way the first one does, so 3 is the guess there, giving us [8, 2, 3]. That's the answer given in the Webb et al. paper.



C. Lewis, *Artificial Psychology*, Synthesis Lectures on Human-Centered Informatics,
https://doi.org/10.1007/978-3-031-76646-6_4

Can a predictive model solve this problem? Yes, as reported by Webb et al., the text-davinci-003 model comes up with this response. A later model, GPT-4, that has been enhanced with fine-tuning, a process by which the model gets some additional training, using example human interactions, actually does worse on this problem.

The complete interaction is shown in Appendix A.2; here are excerpts:

```
User: Complete the following pattern:
[ 5 9 3 ] [ 8 9 2 ] [ 1 9 7 ]
[ 8 4 7 ] [ 1 4 3 ] [ 5 4 2 ]
[ 1 2 2 ] [ 5 2 7 ] [
ChatGPT (using GPT 4.0, 11.10.23):
To complete the pattern, we first need to understand the
rule or relationship governing the numbers in each set.
Let's examine the given sets:
[5 9 3], [8 9 2], [1 9 7]
[8 4 7], [1 4 3], [5 4 2]
[1 2 2], [5 2 7], [...]
Observing each group, it appears that the middle number in
each set remains constant, while the first and last numbers
change. Let's investigate if there's a pattern in these
changes:
In the first row, the middle number is 9. The first numbers
decrease by 3 each time (5 to 8 to 1), and the last numbers
decrease by 1 each time (3 to 2 to 7).
...
The pattern for the first numbers seems to be decreasing
(1, 5, ...), and for the last numbers, it's increasing
(2, 7, ...).
...
Thus, the next set in the pattern could be [9 2 12]. Let's
verify this by checking if it fits well with the overall
pattern.
...
Therefore, the completed pattern for the third row is
[1 2 2], [5 2 7], [9 2 12].
...
```

We can see that this model is trying some sophisticated reasoning, but getting things wrong. For example, in the third paragraph of GPT's response, the sequence (5 to 8 to 1 ) is *not* decreasing by 1.

How can a less advanced GPT 3 model succeed, where GPT-4 fails? The answer is that GPT-4 is approaching the problem the way a human might, by searching for patterns in the numbers that can be described in a simple way. It's using, or at least mentioning, a variety of concepts, like "remaining constant", "trend", and various arithmetic operations. It appears that humans do something similar.

But when Melanie Mitchell, a leading researcher on analogical reasoning, looked at what early LLMs were doing in two 2022 posts Mitchell (2020a, b), she suggested that they were doing something different: "it's hard to tell if GPT-3 is actually making an analogy, or if it is copying patterns in the target strings (Mitchell 2020a, Conclusions)". That is, rather than identifying the *concepts* that a human would think are involved in a problem, Mitchell guessed that the early LLMs were in some way just copying the *structure* of earlier parts of a problem to form a solution.

There's good reason to think that Mitchell was right. One line of evidence is that some of the verbiage that would be used when presenting problems to humans is not needed by simple LLMs. For example, here is how one of Mitchell's examples was presented:

```
Q: If a b c changes to a b d, what does i i j j k k change
to? A: i i j j l l
Q: If a b c changes to a b d, what does m m n n o o change
to?
```

GPT-3 davinci, an early form of GPT-3 with only prediction training, responds as intended, with m m n n p p. But if we remove the explanatory language and present just the examples that are included in the item,

```
Q a b c A a b d Q i i j j k k A i i j j l l Q m m n n o o A
```

GPT-3 still responds as expected about as often as to the original form. The surrounding language is not needed. Just the patterning in the examples is enough.

A second line of argument is based on a kind of reverse engineering exercise. I collected a large number of examples of letter sequence continuation problems that GPT-3 solved correctly, and wrote a conventional computer program that could approximate GPT-3's behavior on these problems.

The starting point for the program, called LoAn, for Low level Analogy, was an observation about the transformer architecture that is used in GPT-3 and other LLMs. As we'll discuss further in Chap. 26, the heart of a transformer is a structure called an *attention head*, that collects information from earlier portions of a sequence, when trying to predict what comes next. In an attention head, a portion of the neural net that is predicting what follows a given item can issue a *query* that is matched to *keys* offered by each earlier token in the sequence. Each key is associated with a *value*, and the result of the query is a sum of all the values, weighted by the similarity of each key to the query. In a 2021 paper, Elhage et al.

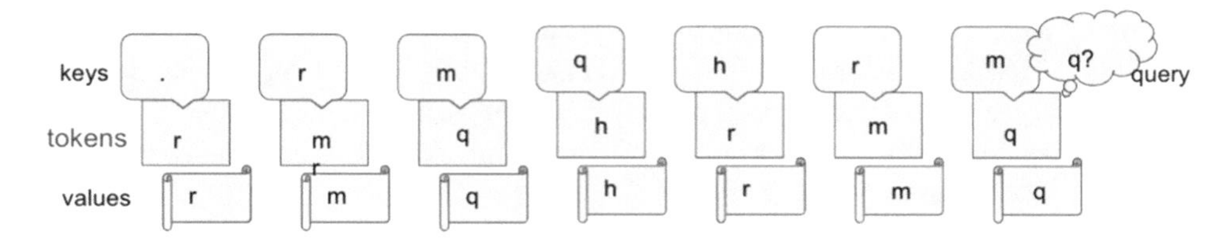


**Fig. 4.1** Shows a sequence of tokens, with keys and values. See text for explanation

(2021); see also Olsson et al. (2022) reported that they often found this machinery being organized in a particular way, that they called an *induction head*.

In an induction head the query is the current token, the value offered by each earlier token is that token itself, and the associated key is the previous token in the sequence, as shown in Fig. 4.1. The effect of this is that if the system is trying to predict what token follows (for example) q, it looks back to see what tokens have been preceded by a q, earlier in the sequence. In the example in the Figure, the query (q) is a good match to only one of the keys, and so the value will be dominated by the value associated with that key, which is h. We can see that that's a reasonable guess, since h followed q earlier in the input.

The logic of this kind of prediction led me to explore a simple generalization: if a *sequence* of tokens, like r m x d, is then partially repeated, as in r m x d r m x, would GPT-3 predict d? The answer is yes, and longer repeated sequences lead to more reliable predictions.

A different line of thinking led to some more complex examples. Experiments with LLMs writing program code suggested that they were, in effect, solving some complicated analogies in that domain.

A use case I was studying (a real-world example) was the creation of a program that would help a brain injury survivor keep track of what gifts she had given to people on her gift list. The program needed to be able to look up a name in a list, and return any recorded gifts. I noticed (a) that the LLM could easily do this, using an appropriate pattern of code to accomplish the task; (b) that the code handled a situation that I had said nothing about in my request to the LLM, in which a name was looked up, but no gifts were recorded; and (c) that the code produced for that situation included a perfectly appropriate message, "No gift found for Fred", with the appropriate name inserted.

How can it do this? While the LLM's corpus very likely includes many example programs that involve looking things up, it seems very *un*likely that any of them addressed looking up gifts, specifically.

Thus, we can suppose that the LLM has responded to a code pattern that is appropriate to this kind of lookup task, and that the code pattern includes the "item not found" case. The LLM then has to modify the "item not found" code to include "gift" as the description of what was not found.

So, how can it interrupt the production of a sequence of code, modify just part of it, to specify a message tailored to this novel application, and then carry on with the rest of the code?

This challenge led me to explore a kind of abstract version of this coding task, cast as a letter sequence continuation problem:

```
[sn]{nnnkss}[cm]{mmmkcc}[yd]{dddkyy}[fq]{
```

We could describe the task here as taking in two letters, in square brackets, as an "input", and producing an "output" in curly braces. To produce the output the system has to produce three copies of the second "input" letter, while remembering the first letter, then insert a "k", and then add two copies of the first letter. In the example, the "input" letters are f and q, so the intended response is three q's, a k, then two f's, and finally a closing curly brace. That's what GPT-3 produces.

There is some fixed structure here, including the braces and the fixed "k", but other parts of the examples have to be modified to use the particular letter pair that is provided as "input". At a very high level, this is a bit like the challenge of writing a piece of programming code for a lookup operation, as in the gift example, while modifying the code as needed to fit a specific application.

The LoAn program can perform this letter sequence task, with no representation of any concepts that might seem to be involved, like making copies, and without "remembering" anything. It also performs as expected on repeated sequence problems, like r m x d r m x.

How does it do this? At a high level, LoAn uses the structure of the earlier part of a sequence as a guide for constructing a continuation, just as Mitchell suggested. One can think of the first part of a sequence as providing a *template* for the continuation.

For the sequence r m x d r m x the templating is very simple. The beginning of the sequence suggests directly that r m x d is a suitable template, that matches the end of the sequence, and need not be modified at all.

The brace sequence is considerably more complex but can still be solved by a generalization of the templating process. The details are presented in Appendix A.2.

The Appendix also shows how the templating process can solve the grid example, as well. The template structure emerges from aligning a sequence with itself, and using correspondences that emerge in an alignment to fill in variable parts of a templated value. Thus LoAn solves problems like these by copying structure from the early parts of a prompt to the end, not by using "concepts", just as Mitchell suggested.

The fact that LoAn solves this problem in this way does not mean that GPT-3 solves it this way. But it does establish that the problem *can* be solved this way, and the operations needed by LoAn seem well within the capacity of the transformer architecture.

Let's return to the performance of various LLMs on this problem. As we've seen, GPT-3 solves it correctly, but GPT-4 does not. Bard also fails on the problem. It imagines that the numbers are part of a solved Sudoku game, and it uses ideas about that game to predict the solution 9, 3, 6. (The fact that the numbers in the grid couldn't legally be part of a Sudoku solution, since there are repeated digits within the rows and columns, does not deter Bard.) So, both of these more sophisticated systems fail on a problem that, as we've seen, can be solved by a simple pattern matching method, and is solved by earlier versions of GPT. The more advanced systems are given fine tuning. But the fine tuning seems to prevent their solving the problem in a simple way.

It appears that we see in the behavior of the different versions of GPT two ways of responding, one using simple pattern matching, and one using more complex step by step reasoning.

That's reminiscent of a distinction in Daniel Kahneman's popular book, *Thinking Fast and Slow* (Kahneman 2011). Kahneman gives many examples of situations in which people make slow, deliberate decisions, and others in which they respond quickly.

Why might slow, deliberate reasoning sometimes fail, when fast pattern matching can succeed? On problems of this kind there are many different relationships that can be identified and explored, separately, most of which are useless. Simple pattern matching can cut through the clutter, while a deliberate process can get caught up in it.

## 4.1 Verbal Analogies

The analogies we've considered so far are deliberately more or less knowledge free, in that the letters and digits that are used have very little meaning attached to them. But the analogies considered by Webb et al., on which GPT performed quite well, include other kinds, for which background knowledge is very much needed.

For example, Webb et al. evaluated GPT-3 on four-term verbal analogies, like these multiple choice items:

```
vegetable : cabbage :: insect : ? A. beetle B. frog
love : hate :: rich : ? A. poor B. wealthy
drive : car :: burn : ? A. wood B. fire
rob : steal :: cry : ? A. weep B. laugh
```

Plainly one has to know the meanings of these words, in order to answer the questions. A pure pattern matching system like LoAn is helpless on this kind of material.

These are multiple choice items for which a purely predictive model like GPT-3 davinci will struggle because the needed response is not simply some text that might follow the prompts in the corpus. (Webb et al. evaluated GPT not from its responses, but from an internal measure of probability that they obtained for the different candidate continuations.) But if we don't provide choices, these verbal analogies can be presented in a form that only calls for a continuation, for example

```
love : hate :: rich :
```

The purely predictive davinci model gets the intended answer: poor. It also produces the intended response, "weep", for

```
rob : steal :: cry
```

For

```
vegetable : cabbage :: insect :
```

its response is "bug", which while not the choice offered in the multiple choice version of the item that Webb and collaborators used, certainly seems appropriate. For

```
drive : car :: burn :
```

its response is "fire", some way off the intended track: "wood". But for that item we can see that a proper response is difficult to discern, without a choice of alternatives.

These items, as presented, may take advantage of things that GPT "knows", beyond these word meanings. The delimiters : and :: are conventionally used when presenting analogies; maybe GPT recognizes them? What happens if we replace those with arbitrary letters?

```
love r hate z rich r
```

gives the response "poor", as before. So the familiar (to people) delimiters may not be helping GPT.

A more comprehensive investigation of the antonym problems, like rich : poor, can be done. First, we can collect all of the three term analogy items that can be generated from this list of 14 antonym pairs:

```
good bad
rich poor
up down
hot cold
happy sad
ugly beautiful
young old
left right
wet dry
black white
true false
love hate
male female
in out
```

For example, from the pairs good, bad, and rich poor, we can form the eight items

```
good: bad :: rich :
bad : good :: rich :
good : bad : poor :
bad : good : poor :
rich : poor :: good :
poor : rich :: good :
rich : poor :: bad :
poor : rich :: bad :
```

In this way, the 14 pairs produce a total of 728 items (not all independent, of course, since any of the pairs occurs in 52 items as the first part of the item, and in 26 more items as the final part.)

Using this collection of items, we can compare different presentations of them. In analyzing the results, any response that can reasonably be considered an antonym of the third term is scored as such, not just the item paired with it in the original list. For example, "pretty" is accepted as an antonym of "ugly", even though the word paired with "ugly" in the list of antonym pairs is "beautiful".

The standard delimiters, : and ::, perform quite well on this test, overall 80% correct. On the other hand, using the arbitrary letters r and z as delimiters, so that items look like

```
love r hate z rich r
```

do quite poorly, producing antonyms on only 10% of the items.

Does this mean that there is indeed some special virtue in the conventional delimiters? Not in the obvious way; testing also shows that the delimiters - and *, in which items are presented as

```
hot - cold * rich,
```

actually perform slightly better than the conventional : and ::, at 82% correct.

Thinking back to the LoAn model, discussed earlier for meaningless letter strings, we saw that the model used repeated patterns in a prompt to produce its predictions. From this point of view, we might think that the prompts we are using for these word analogies are problematic: because the double colon delimiter appears only once in each item, it can't participate in any repeated pattern in the prompt. What would happen if we created a repeated pattern, by putting an extra :: at the start of each item, as in

```
:: hot : cold :: rich :?
```

This leads to a substantial increase in accuracy. On the complete set of 728 items, this form yields 99% correct responses.

But is the repeated :: really responsible for the improvement? If we replace the leading :: with an asterisk, so items look like

```
* hot : cold :: rich :
```

we still get 98% accuracy!

Are delimiters needed at all? If we eliminate them completely, so that items look like

```
hot cold rich
```

we get only 9% antonyms. Putting :: at the beginning of each item gives 58% antonyms, which is an improvement, but not at the same level as the items with the common delimiter pattern. We can see that the delimiters do matter, but it's not clear how. Presumably, the way specific delimiters, and patterns of delimiters, occur in the training corpus, plays some role.

However, it is possible to get a high yield of antonyms, with no use of delimiters at all. By constructing items that consist of a *series* of antonym pairs, followed by a single word, GPT will predict an antonym much of the time. Figure 4.2 shows the results for sequences of 1–10 antonym pairs, with no delimiters, using samples of 500 items constructed from the same list of antonym pairs we have been using (The data are also shown in Appendix A.2, Table D.1). We'll discuss other data in the figure shortly.

Here we can see further evidence that GPT is not reliant solely on recognizing the conventional presentation of analogy items, since these items use no delimiters at all. On the other hand, the overall pattern of results, with the conventional form being slightly better than forms using other delimiters, and single pairs with delimiters yielding as many antonyms as sequences of multiple pairs without delimiters, does suggest that delimiters play some role.

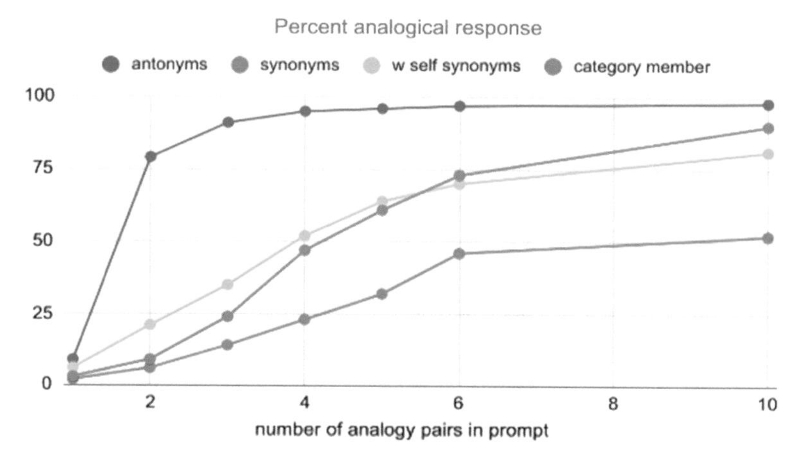


**Fig. 4.2** Prompts with multiple analogy pairs

### 4.1.1 Synonyms

What about other types of word analogies? Using a list of 15 synonym pairs, shown in Table 4.1, we can do the same kind of tests as for antonyms. Using the delimiters ::, :, ::, :, the form that performed best for antonyms, we obtain only 70% synonyms. Results of tests with varying numbers of pairs of synonyms, followed by one member of a synonym pair, with no delimiters, are shown in Fig. 4.2. The figure also shows what happens for antonyms, as we've seen, and for category-member pairs, that we'll discuss in a bit.

Why does GPT perform worse on synonyms than antonyms? One issue seems to be producing a word as its own synonym, that is, counting words as self-synonyms. As one of the lines in the figure shows, counting these as correct raises the score, but it's still a good deal short of the yield for antonyms. For the items with 10 pairs of synonyms, the yield rises to 81% for synonyms, while for antonyms it's over 90%. The figure shows the performance for items with smaller numbers of pairs in the prompt, too.

### 4.1.2 Categories and Members

The figure also shows the performance for a third kind of analogy, one that links a category to a category member, like

```
city: London :: disease :
```

where "measles" or "flu" would be appropriate responses. We'll use the pairs shown in Table 4.2.

**Table 4.1** List of synonym pairs

| Simple easy |
|---|
| Mad angry |
| Ill sick |
| Tiny little |
| Stone rock |
| Large big |
| Toss throw |
| Skinny thin |
| Enjoy like |
| Late tardy |
| Giggle laugh |
| Hop jump |
| Loud noisy |
| Brave courageous |

**Table 4.2** List of category-member pairs

| Vegetable cabbage |
|---|
| Fruit apple |
| Mammal rabbit |
| Pet dog |
| Vehicle car |
| Country Peru |
| City London |
| River Nile |
| Clothing sweater |
| Soup minestrone |
| Furniture sofa |
| Disease measles |
| Tree spruce |
| Grain wheat |

When given single items like these, with the :: : :: : delimiters, GPT responds correctly 70% of the time. Figure 4.2 shows the performance for prompts that include varying numbers of these pairs, with no delimiters (The data are also shown in Appendix A.2, Table D.2). For prompts with 10 pairs, 90% correct responses are produced.

## 4.2 Analogical Reasoning . . . Why?

We see that for a variety of analogical reasoning tasks, GPT does quite well. Why is it able to do this? Materials like those seen in the tasks we've discussed, for example the prompt

```
mammal rabbit fruit apple river Nile city London grain wheat
vehicle car clothing sweater furniture sofa vegetable cabbage
tree spruce soup
```

can hardly have appeared in the training corpus. A related question is, why does the system respond "analogically" at all, to items like this, with no task description or instructions? We don't have to say, "Given these word pairs, followed by a single word, what word would complete the pair at the end?" or anything of the sort. We just type in the words, and GPT completes the pair analogically 90% of the time.

We can suggest answers to these questions by relating analogy to the demands of *generalizable prediction*. It's not enough for an LLM to recall what tokens have followed a given sequence of tokens in the training corpus. Rather, it has to attain a *generalizable* prediction ability, the ability to make accurate predictions for sequences it has not seen yet, on the basis of what it has learned from sequences it has seen. That's because, during training, it is constantly being given new parts of its corpus, and if it gets those wrong, its network gets changed.

One approach to this prediction problem is the use of analogy. A general framing of an analogy is as a function f that connects two things:

A : $f(A)$

For example, in structure mapping (Gentner 1983, 2010), the dominant account of human analogical reasoning, A and $f(A)$ may be representations of situations, with f being a mapping that carries parts of A to parts of $f(A)$. In the copycat theory of Mitchell and Hofstadter (1995), A and $f(A)$ are letter strings, with f being a way to "do something" to strings. A task for copycat is, given strings A and $f(A)$ (some transformed version of A), and a second string B, produce $f(B)$:

A : $f(A)$ :: B : ?

In our situation, we can focus on cases in which A and B are sequences of tokens. Now suppose that the training corpus contains the sequence A S, that is, A is followed by some further sequence of tokens, S. If S can be recognized as $f(A)$, for some f, then it is a plausible guess that B, which can be a previously unseen sequence, may be followed by $f(B)$. That is, by detecting f when it occurs with A, we obtain the ability to make a prediction for a sequence we haven't seen before, as desired.

This prediction is fallible. We may encounter a B for which f is just not applicable, as in these examples:

```
1 2 6 7 9 10 4> 
5 1 2 6 7 9 10 d> 1
```

In the first example, GPT-3 davinci is able to detect "add one" as the likely operation that relates the first item in a pair to the second item, leading the prediction of 5 to follow 4. But it can't apply this operation to d, in the second example.

It could also be that f is applicable, but actually B does not predict $f(B)$. But it seems reasonable to consider the various f's that may be detected in the course of training as sources of votes in favor of various possible predictions.

This idea provides a possible answer to the questions we framed earlier: How can an LLM respond analogically to material quite unlike what it has seen in its corpus? And why does it respond analogically, in the absence of any explicit instruction to do so? First, analogical predictions may be valuable for generalizable prediction. Second, if analogies play this role in the prediction process, then no special instructions are needed to evoke analogical responding. Rather, analogical responding is built into the prediction process. That's what we see in simple LLMs, in which simple prediction behavior isn't overridden by fine-tuning.

## 4.3 Analogical Reasoning... How?

Let's suppose we accept that being able to use analogies in this way is valuable in the prediction tasks on which LLMs are trained. The questions now arise, given a sequence A S, how can S be recognized as f(A), for some f? And how can f be applied to a novel sequence B?

A popular account of this uses *vector encodings* and *vector arithmetic*. Inside an LLM entities are represented, or encoded, as vectors, that is sequences of numbers of some set length. In GPT davinci, the length of these vectors is more than 12,000. Roughly, this means that tokens can differ in more than 12,000 independent ways, and vastly more ways than that, by allowing these numbers to vary in combination. Vectors of this same size can be used to represent sequences of tokens, including words.

These vector encodings are created during training. One can think that the encodings that are used are whatever makes the prediction process work as well as possible, based on the feedback the system is given during training.

Vector encodings have been used in many different contexts, and it was observed long ago (Rumelhart and Abrahamson 1973) that they allow some word analogies to be recognized and processed in a simple way. A classic example is man : woman :: king : ?. If these three words are represented by vectors, one can represent the transformation that maps man to woman as the difference between the vector for man and the vector for woman.

If $\mathrm{V}_{\mathrm{man}}$ and $\mathrm{V}_{\mathrm{woman}}$ are the vectors representing man and woman, the difference $V_{woman} - V_{man}$ is another vector, obtained by subtracting each of the numbers in $\mathrm{V}_{\mathrm{man}}$ from the corresponding numbers in $\mathrm{V}_{\mathrm{woman}}$. The effect is that, to transform $\mathrm{V}_{\mathrm{man}}$ into $\mathrm{V}_{\mathrm{woman}}$, one can just add the vector $(V_{woman} - V_{man})$ to $\mathrm{V}_{\mathrm{man}}$.

So we can represent the function that maps man to woman by the operation of adding the difference $V_{woman} - V_{man}$. To solve the analogy, we just apply this operation to the vector for king. Often this vector agrees closely with the vector that represents queen.

This method solves both of the problems noted above. We can easily discover the f that maps one word to another: we just subtract the vector for the first word from the vector for the second word. Then to apply that f to another word, we just add that difference vector to the vector for that word.

It's plausible that this mechanism is used inside GPT to solve some of the analogies we've been discussing. But it is difficult to tell, because it's not easy to determine how a given word is represented inside GPT. On the other hand, we can be *sure* that vector arithmetic is not the whole story.

A clear problem is that analogies involving antonyms and synonyms can't be solved this way. That's because antonyms and synonyms are *symmetric*: if A is an antonym of B, then B is an antonym of A, and the same for synonyms. For example, hot is an antonym of cold, and cold is an antonym of hot. Now consider the two analogy problems

```
hot : cold :: rich : ?
cold : hot :: rich : ?
```

The solution to the first should be $V_{rich} + V_{cold} - V_{hot}$, and the solution to the second should be $V_{rich} + V_{hot} - V_{cold}$. But these answers should be the same! That can only happen if $V_{cold} - V_{hot}$ is zero, which can't be right. If it were, the solution to both analogies would be "rich".

We can follow one of two paths from here in our discussion. One path is to acknowledge a gap in our understanding of LLMs. We can say that the system can identify transformations f, in many situations, in some way, and that it can apply these transformations, in some way. But we don't know how it does either of those things. As we'll see, there are some further observations we can make, even with these gaps.

The second path is to speculate about a different mechanism, not vector arithmetic, that might support the operations of identifying and applying transformations. We'll call it *virtual abduction*. Abduction is a form of reasoning in which, given some observation, one guesses that something that would *explain* the observation is also true. Abduction is fallible, but very often works well: seeing the track of a deer in the snow, we suppose that a deer has passed by. That's because the passage of the deer would explain the track.

The approach to analogy will be to identify operations on sequences of tokens with other sequences of tokens, as follows. Consider the sequence of tokens that represents "give the opposite of": let's abbreviate this as OPP. If we follow OPP with hot, the system will predict cold. We can think of OPP as representing the f that maps hot to cold. Now consider the sequence

```
hot cold rich
```

The system can notice that cold has followed hot. What could explain that? It would be expected if hot had been preceded by OPP. So the system can use abduction, and operate as if OPP is in the context. It's not really there, of course, so this is only *virtual* abduction. With OPP in the context, rich will tend to predict its opposite, which is poor.

So in virtual abduction, the functions f are represented by sequences of tokens, and they are determined by asking what sequence of tokens would explain the observed sequence, in which cold follows hot. The f is applied by asking, what is predicted when OPP is followed by rich.

In the example of antonyms we can imagine that there could be a definite sequence, "give the opposite of" that can do the work required of OPP. But this idea doesn't hold up. There are indefinitely many sequences that could play that role, such as "give an antonym of". As we've seen, there are sequences that don't include any abstract term like "opposite" or "antonym". Indeed, many of the sequences of antonym pairs that were discussed earlier, like

```
up down wet dry bad good
```

do a pretty good job of acting like OPP.

This variety might be accommodated by imagining all of the candidate sequences, being virtually present to some extent, with a strength or weight determined by how often they have made the correct prediction during training. The strength or weight is a number that determines how much or little the given sequence contributes in determining a prediction, in a kind of voting process. The weight of a sequence may also be affected by how recently it has been evoked, with its importance decaying over time.

The voting process is the way competing predictions are adjudicated. Looking at the internal machinery of GPT, aspects of a given context will assign probabilities to every possible token, of which there are about a thousand. These probabilities are added up, and the token with the largest accumulation of probabilities is the one predicted.

With this background, we can look back at the curves in Fig. 4.2. There we see that a single antonym pair is rarely effective in producing an analogical response. Two pairs are more effective, three still more so, and so on.

This effect might be explained as the result of the abduction process creating virtual contexts that predict antonyms, pushing the prediction in that direction. As the sequence of pairs is extended, the influences add up.

We can add some additional observations to the picture. Consider this prompt:

```
clothing sweater soup minestrone pet dog tree spruce
vegetable cabbage furniture sofa fruit apple disease
measles grain wheat vehicle car country
```

This includes 10 pairs, each consisting of a category and a category member. As we saw earlier, GPT generally responds analogically to prompts like this, and it does here: it responds with the name of a country: "Italy". But why Italy? A little experimentation suggests that the reason is the presence of the pair "soup minestrone" in the prompt. Minestrone is an Italian soup. This influence, with "minestrone" pushing for a specific choice of country, suggests a way to probe the considerations that shape GPT's responding to prompts of this kind.

We can create some background by showing that the following three prompts are reliably responded to with the appropriate country names:

```
city London river Thames country
city Paris river Seine country
city Moscow river Volga country
```

We can now ask, what happens when we interpose some *filler material* between the city and river pairs and the final item, country? Abductive analogy would suggest that each pair would create a virtual context in which (in the first case) city would predict London, and then river would predict Thames. In this context, it seems likely that country would predict England, or perhaps United Kingdom. But would the context remain effective if we insert material between Thames and country?

Figure 4.3 shows what happens when we insert increasing numbers of antonym pairs in these prompts. As we know, when there are no antonym pairs inserted, the system always pre-

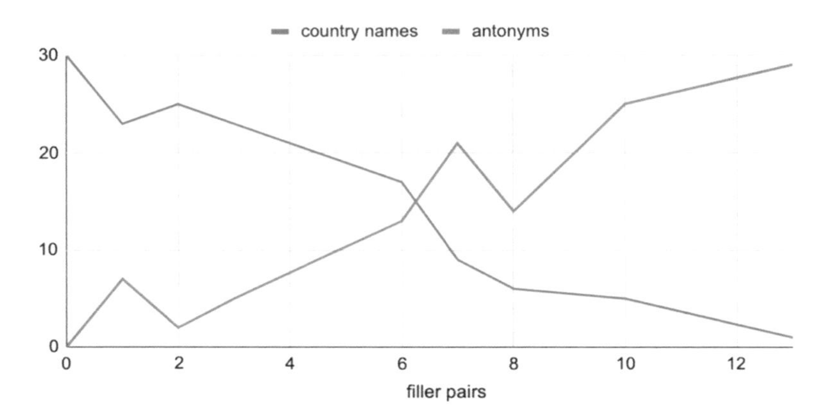


**Fig. 4.3** Responses to prompts with filler pairs (among 30 items)

dicts England (or United Kingdom), France, and Russia for these three prompts. But as more and more antonym pairs are inserted, the responses change. The antonym pairs are creating the virtual context OPP, that we discussed before, so votes are being split between an antonym of "country" ("city" or "town"), and the country name. When we have inserted 13 antonym pairs we get only 1 country name and 29 antonyms of country ("city" or "town"), among 10 versions of each of the three prompts, with a different random selection of antonym pairs.

Figure 4.4 shows another assessment of the impact of city London river Thames (and corresponding material for France and Russia). There we are inserting not antonym pairs, but category-member pairs, like grain wheat. (The data are shown in Appendix A.2, Table D.3.) The inserted pairs don't include any city, river, or country pairs, and the soup pair has been

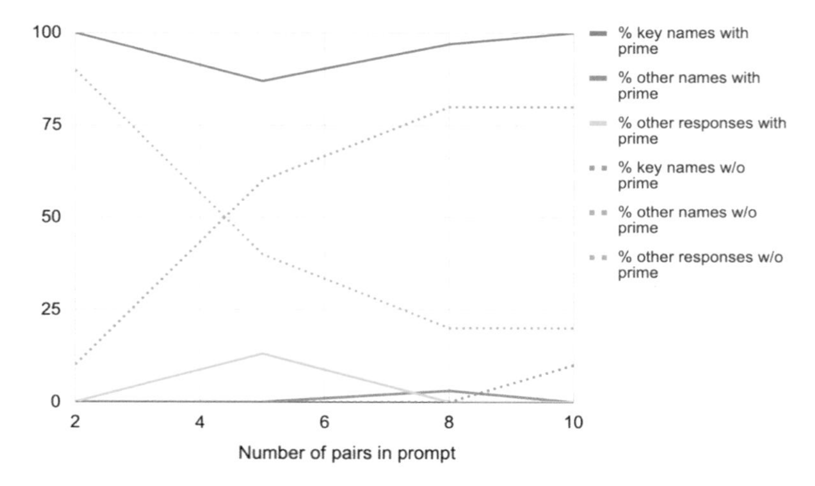


**Fig. 4.4** Responses to prompts with multiple category-member pairs

changed to feature chicken rather than minestrone, to avoid the strong association between minestrone and Italy.

Let's focus first on the solid lines in the figure. They show that, as we insert more and more category-member pairs, nearly all the predictions are the names of the key countries, England/United Kingdom, France, or Russia. Only a few predictions are for anything else, among 30 trials overall for each number of filler pairs. Note that the x-axis shows the total number of pairs in each prompt, including the "priming" city and river pairs.

The fact that the proportion of key country name responses stays high, as more and more filler material is inserted, suggests that the virtual context established by the city and river pairs is not decaying, at least over the range of insertions studied. The fact that the effectiveness of the context did diminish in the trials shown in Fig. 4.3 would therefore not be attributed to decay, but to interference. That is, the inserted material reduces the effect of the earlier context not by delaying the test, but by introducing a different context, OPP, whose votes compete with, and come to dominate, the votes for the key country names.

The dashed lines in the figure show what happens when the "priming" city and river names are not supplied. As we saw earlier, as the number of category-member pairs increases, the number of category-member predictions increases, in this case, predictions that "country" will be followed by the name of a country. But without the city and river material biasing the predictions towards the key countries, those countries are almost never predicted, even though more and more of the predictions are of country names. In fact, a key country name showed up only once, among 10 trials with random category-member preceding the final "country" prompt.

Summarizing, these observations suggest that the virtual contexts evoked by pairs in a sequence can compete, reinforce one another, or blend. We saw competition between the contexts evoked by antonym pair and category-member pairs. We saw reinforcement, when increasing numbers of pairs that reflect a given relationship are presented. We saw blending, when the city and river pairs shaped the particular choice of country, when making a prediction of an item to follow the country prompt.

With this possible broad brush picture in mind, we can rejoin the thread from the gap in our knowledge of how analogy actually works. Perhaps virtual abduction doesn't work, or can't work, or could work, but isn't how LLMs, or humans, accomplish analogies. Sadly, we just don't know how LLMs do analogical reasoning. But without committing ourselves to particular ideas about mechanism, we can move forward with the idea that material in a prompt that instantiates a particular relationship promotes predictions that also instantiate that relationship.

This means that the system is able to identify a relationship that is instantiated in a prompt, and that it is able to make predictions that instantiate that relationship, even if novel material is involved.

Why would this happen? Because, as we have discussed, doing this during training is a way to make predictions in contexts that have not been seen previously in training. After training, it supports making reasonable predictions for material that is not in the corpus at all.

The paper by Webb and collaborators includes many other kinds of analogies, including between stories, on which GPT does well. We saw an example of this kind of thing earlier, when GPT was able to compare Gerhard Fischer's two problems. It was able to do this, even though the settings for the two problems were totally different: chessboard and marriages. But we're having enough trouble working out how even much simpler analogies are being processed.

## 4.4 Structure Mapping

As mentioned earlier, the dominant theory of analogical reasoning is structure mapping (Gentner 1983), which involves building structured representations, like trees, for two situations, and building connections between them. As far as we can tell, nothing like that is happening in GPT. Certainly, nothing of the kind was put into it by its designers. We'll discuss later on the possibility that nevertheless structures *emerge* during training. But if they don't, the success of LLMs on this kind of task suggests a big departure from current thinking.

## 4.5 Taking Stock

The key messages here are:

- Analogical reasoning is finding relationships between situations, and applying those relationships to what's known about one situation so as to answer questions about another situation.
- Existing LLMs do quite well on analogical reasoning, and hence it's reasonable to think that PR could, too.
- LLMs do this with no explicit provision for mental structures, that theories of human analogical reasoning have assumed necessary. This is an instance of a recurring theme, that the success of LLMs casts doubt on the importance of structures in many areas of psychology. Their success also motivates further investigations of how LLMs do what they do.
- Analogical reasoning in LLMs arises from training on a vast amount of text, with no explicit emphasis on analogical reasoning as such. That is, the LLM receives no instruction of the form, "To solve an analogy, this is what you need to do". Thus, it seems that analogical reasoning is needed to be able to make successful predictions in text, and it is learned for that reason.
- We know that LLMs can do analogical reasoning, and we can suggest *why* they can do it, but we can only speculate about *how* they can do it. Again, more work is needed on this.

In the chapters to come, we'll have occasion to apply PR's analogical reasoning in a number of settings, for example how and when PR would imitate behaviors that is observes, or how it can interpret gestures. There are remaining mysteries about *how* analogical reasoning is done, but we can move ahead with what we've learned: LLMs *can* do analogical reasoning, and often *do* do it, with no explicit instruction.

## Notes

(1) If analogies are intriguing for you, you'll enjoy the wonderful *Surfaces and essences: Analogy as the fuel and fire of thinking*, by Hofstadter and Sander (2013).

(2) Another mental process that's been thought of as involving structures is *conceptual blending*. That's the operation of combining aspects of two conceptual structures to form a new one. An example from one of the early presentations, (Fauconnier and Turner 2003), is a description of a "race" between a contemporary sailing vessel, and a vessel of an earlier century. The idea of the race was that the contemporary vessel would follow the same route as its predecessor, and it would win if it took less time to complete the voyage. As a blend, the authors suggest that this "race" is formed by combining aspects of the contemporary voyage with aspects of the earlier one.

A simpler example, also from Fauconnier and Turner (2003), is a family of constructions, "dolphin-safe", "shark-safe", and "child-safe". As blends, these are formed by combining ideas about dolphins, sharks, and children, with ideas about danger.

The way these ideas are combined depends on how these ideas relate to one another, and to the larger context, ideas about danger. Thus "child-safe" means safe *for* children, since we think of children as needing protection from danger, while "shark-safe" means safe *from* sharks, as sharks aren't seen as experiencing danger, but creating it. The blend idea has been a fruitful one, and examples have been cited in many fields.

Turner and other researchers describe blending as an operation on "structures", as I've done in introducing the idea. These structures are generally described as networks, and blending as an operation on these:

> Blending work can be done anywhere in the blending network at any instant in the construction of the network, including in developing various spaces and developing vital conceptual relations between them. ... Blending networks recruit from stable structures and can themselves become stable, widely shared, and entrenched. (Turner 2020, P 2)

Talk of networks and structures comes naturally in current psychological and linguistic theory. But can blending happen without these things?

One of the very first descriptions of the GPT LLMs (Mayne n.d.) included a conceptual blend. The system (GPT-3, at the time) responded to this prompt

```
1. Batman 2. Superman 3. Cyborg
```

this way:

```
4. Flash 5. Martian Manhunter 6. Green Arrow 7. Hawkman
```

Readers like me, not well up on US popular culture, will want a bit of explanation. All of the entries in this list are names of comic book heroes, and, further, are heroes in comic books issued by the same publisher, DC Comics. We can see that GPT has identified the names in the prompt as that kind of entity, and has populated the rest of the list with entries of the same kind.

Our interest in the example isn't that, however. It's that GPT has *blended* the concept of DC Comics heroes with the concept of numbered list.

We know how that blending happens. GPT's prediction process, with its voting mechanism, weighs the regularities that connect the list entries with the regularities that define a numbered list, and emits the appropriate tokens, in the order seen in the response. We don't have to imagine structures in any of that.

# More Reasoning: Inference and Prediction 5

***Foci***: Brandom (2001) *Articulating reasons: An introduction to inferentialism*, and Harman (1986) *Change in view: Principles of reasoning*. We often think of reasoning as a sequence of logical operations, in which some ideas imply others. Is that a good account of reasoning? Can the predictive perspective offer any insights on this?

The rules of logic are often taken to be principles of proper reasoning. For example, from a proposition P, and the implication P implies Q, one can infer Q. This and other rules do indeed capture *soundness*, with a technical meaning: a *sound* argument is one for which, if its premises are true (P, and P implies Q, in this case), then its conclusion (Q) is guaranteed to be true. Some thinkers have proposed that logical rules of this kind are fundamental to the process of thought.

The rules of logic do tell us about soundness, but there are good reasons to doubt how much the rules of logic actually tell us about thinking. The philosopher Brandom (2001) notes that a great deal of reasoning (nearly all?) doesn't deal with abstract patterns like the one in the example, but rather relationships between concrete situations. For example, from "this is a match", and "I will strike it", I may infer "this match will light", with no logical connectives appearing anywhere.

One might think one could just rewrite these ordinary sentences to expose an underlying logical relationship. This could give something like

```
if X is a match, and X is struck, then X will light
```

But Brandom points out that this is far too simple. There is an indefinite list of conditions that would have to be included in the if part of a statement like this to make it true: the match can't be damp, it has to be in an atmosphere that contains enough oxygen, there can't be too



C. Lewis, *Artificial Psychology*, Synthesis Lectures on Human-Centered Informatics,
https://doi.org/10.1007/978-3-031-76646-6_5

strong a magnetic field, and so on. Translating a simple, everyday piece of reasoning into a logical implication can't really be done, in practice, because of the indefiniteness of that list of considerations.

A second issue is pointed out by another philosopher, Harman (1986). The rules of logic at best tell you what you *can* infer from some premises, but not what you actually *should* infer, in some situation. For example, from the premise "P implies Q" one can infer any of "P and R implies Q", "P and S implies Q", "P and T implies Q", and so on. But the rules of logic don't tell you which, if any, of those inferences you actually want to draw.

Further, Harman points out that sometimes, when you think P, and "P implies Q", you don't always want to infer Q. If you think that Q is absurd, what you want to do is cancel your belief in one or more of P, "P implies Q", and "Q is absurd". Inferring Q is only one of the possibilities.

A concern that both Brandom and Harman point out is that logic is what is called *monotonic*, whereas thinking commonly is not. In logic, if one adds something to one's beliefs, the number of implications, other things one should believe, always increases. But very often adding a new belief means that you retract some current belief. The situation just described can provide an example: if I believe "P implies Q", and I believe that Q is absurd, if I now become convinced that P, I have to revise my beliefs by abandoning my belief that P, or that "P implies Q".

Viewing inference as the operation of predictive processes can help with these issues. Let's take the following schema to represent an inference:

```
I am thinking about A, B, and C. Next I should think
about D.
```

Here the thoughts A, B, and so on can be premises, but also thoughts that represent goals. The schema expresses the fact that I should think whatever I should think next, given those current or recent thoughts.

In a prediction framework, the schema would be expressed as a prediction. If I'm thinking these things, the schema votes for the prediction that I'll think about D next. However, there will also be other aspects of the model that will vote for other things, and the actual prediction, that is, what I will actually think of next, will be determined by the voting process.

This setup can address the weaknesses in the logical view. First, in the match problem, the indefinite collection of counterindications for a match lighting will be represented by votes for predictions that the match won't light. (Actually, only the counterindications that I know about will enter in that way. If I don't know about the magnetic field condition on lighting, or don't know that a strong magnetic field is present, I might expect that a match would light, and be wrong.)

The counterindications don't have to be organized into a single expression, as using simple logic would require. The experience relevant to expecting things about struck matches can be built up incrementally.

Harman's requirement is also met in the prediction account. My expectations could conceivably include lots of irrelevant logical possibilities, like D or E, D or F, and so on. But these would be outvoted by predictions more responsive to the actual context in which the inference is being made. The context will include what I am trying to do, in many situations.

Does representing "'what I am trying to do" require some formal apparatus? No. Any thought that influences how my reasoning will be expected to evolve will suffice.

The same voting logic means that reasoning, thought of as prediction, isn't monotonic. That is, adding something to the context in which a prediction is made can change the result. Thus, a thought that would have occurred in one context might *not* occur in a modified context.

Brandom's discussion of the match problem is part of a larger program called *inferentialism*. There he argues that the most important aspects of meaning in many situations are what follows from some statement, and what it follows from, not what the statement refers to. He also suggests that logic has arisen from our efforts to describe the patterns of reasoning we observe in ourselves. Thus logic isn't fundamental to thought, but is rather a kind of commentary of some patterns in thought. The prediction framework seems broadly compatible with Brandom's analysis.

Brandom's match example may shed some light on why LLMs often seem to us to be "intelligent" rather than just "knowledgeable". Their training on vast corpora might seem to suggest the latter characteristic, "knowing" a lot, not the former. But the match example shows that thinking effectively about things (being "intelligent") can rest not just on enacting patterns of thought, but also on knowing things, such as that strong magnetic fields affect flames.

## Note

The discussion of "logic" in this chapter addresses only one, traditional system, and there are many others. For example, while traditional logic is monotonic, as discussed, so that adding a belief always leads to more possible inferences, there are various non-monotonic logics, that support different kinds of inference. See Strasser and Antonelli (2024), which includes some discussion of non-monotonic logic as providing insight into human reasoning.

# 6 Transfer of Skills, Production Rules, and Prediction

> ***Focus***: Polson et al. (1986) A test of a common elements theory of transfer. For people, learning one thing can make it easier to learn another thing. This has suggested ideas about how knowledge is represented in memory. How would this work for PR?

Sometimes, when we have to learn to do something new, we can profit from already knowing how to do some other, related thing. For example, when I am driving a rental car, the heater and sound system controls may be quite different from my car, or from other rental cars I have driven, but I will usually feel that I can use what I know about other cars to help me learn about this car. Psychologists call this *transfer of skills*, or *transfer of learning*, or *transfer of training*, or just *transfer*. The image is that knowledge acquired in one situation can be moved over into a new situation, where it provides some value.

One way to detect transfer is to have some people learn to perform some task, say B, after having learned to perform another task, say A, while other people just learn to perform task B. If there is transfer from task A to task B, then the people who have already learned to perform task A, should learn task B more quickly than people who started in on B without learning A.

Polson et al. (1986) used this method to study transfer between text editing tasks, things like changing the line spacing in a document, or printing a document. (Text editors were different in 1986 from today, so some of the tasks would seem strange to us, like loading a document from a floppy diskette. Does anyone even remember floppy diskettes?) The researchers asked their participants to learn various tasks in various orders, and they measured the time needed for the participants to complete an automated training program that required them to perform each task three times with no mistakes before moving on to other tasks.

To analyze the results, the researchers developed a production system model for the whole collection of tasks. This has become a common way to represent knowledge of how to do



C. Lewis, *Artificial Psychology*, Synthesis Lectures on Human-Centered Informatics,
https://doi.org/10.1007/978-3-031-76646-6_6

things, used in a great deal of theoretical and applied research. One can think of a production system as a kind of computer program, only being executed in a person's head, not on a computer.

A production system is a collection of *production rules*, each of which consists of a *condition* and an *action*, and some collection of information, called *memory*, on which the production rules act (in a way that we'll explain). The researchers don't give any examples of production rules in the paper we are focusing on, but in other work they do. Here are a couple (from Bovair et al. 1990), translated into English from a computer notation:

```
Condition: the current goal is ''copy string'', and the
current step is ''verify copy''
Action: verify the copy (by looking at the screen), delete
the current step and add the step ''press accept'' as the
current step
Condition: the current goal is ''copy string'', and the
current step is ''press accept''
Action: press the ''accept'' key, delete the current step,
and make the current step ''finish up''
```

To follow what's happening here, imagine that a person is performing a task that involves copying a string that they see on a computer screen. At the moment, imagine that they have formed the goal of copying the string, and they have already done whatever was needed to select the string to copy, and to make the copy. To finish up, they need to check that the copy has worked properly. In their memory we'd see the goal of copying the string, since they haven't finished with that yet, and an indication that the step they are on is the verification step.

The condition of the first rule looks to see if the indicated information is in memory. We're assuming that it is, so that condition is satisfied. When the condition of a production rule is satisfied, its action is carried out. In our example, that means looking at the screen to check, and also changing what's in memory: it takes away the old current step and adds a new one, to press accept.

With that change in memory, the condition of the first rule is no longer satisfied. But the condition of the second rule now is satisfied, and so its action is carried out. That means pressing the accept key, and changing the current step to indicate what has to be done next. Other production rules would carry forward the work from here.

None of the details of the example matter, other than in providing a concrete idea of what production rules are like. They have conditions that test things in memory, and actions that do things when the tests are satisfied. Crucially, actions can change what is in memory, so that, as we see in the example, the action of one rule has the effect of enabling another rule to work.

Returning to transfer, and how Polson et al. evaluated it, the researchers created production rules for all of the operations required in their collection of text editing operations, including any necessary management of what operations to perform when, in carrying out a task.

Having done that, they could examine any given task, and determine just which productions are needed to perform that task. Armed with that analysis, and some strong assumptions, they then made a whole set of predictions about how long it would take people to learn a given task, having already learned, or not learned, some other tasks. That is, they could predict transfer between learning one task and learning another, by examining the production rules needed to perform each task.

Specifically, they assumed:

1. To perform some task, say task B, one needs to know some particular list of production rules.
2. If one doesn't already know a production rule, it takes some definite amount of time to learn it. They estimated this time from their data to be about 25 s.
3. If one already knows a production rule, because it is needed for some other task, say Task A, and one has already learned task A, it takes no time at all to learn that production rule.

These assumptions, and their lists of what production rules are needed for each task, allowed the researchers to predict how long it would take a participant to learn to perform any given task, depending on exactly which tasks they had already learned.

For example, one of the tasks was to load a diskette and make a copy of it. One group of participants had to learn this task first, without learning any other tasks. The researchers' analysis said that these participants had to learn 52 new rules to perform that task. Another group of participants learned to perform four other tasks before learning that one. They would have to learn only 25 new rules because all the other rules had to be learned already.

What did the data show? For these two groups, the times to learn the duplicate task were 1600 s for the group that had to learn it first, but only 900 s for the group that had learned other tasks first. That difference equates to the time needed to learn 28 new rules, very close to the difference of 26 rules predicted by the researchers' analysis. Overall, the researchers' predictions, based on their production system analysis, proved quite accurate.

An enormous amount of other research has established that production rules are a useful way to describe what must be learned to perform a task. To cite just one example, Taatgen et al. (2008) show that it's possible to predict the time required to learn a family of complex operations on a simplified flight management computer. Production systems are a key element of an extremely successful, very comprehensive theoretical program in cognitive psychology, John Anderson's ACT-R project (for a review see Ritter et al. 2019).

Does this success mean that PR, as a model of learning and action is a nonstarter? On the face of things, it may seem as if a predictive model, like PR, has nothing in common with production systems. Hence good predictions based on production system analysis, like those in the Polson et al. paper we've focused on, and many, many other papers, count as evidence against predictive models.

A closer look suggests a different conclusion, however. The action of a production rule specifies actions, like pressing a key in the example, or a change in memory. The condition specifies what has to be in memory for the action to be performed. If we view memory as the context in which things happen in the production system, we can view each production rule as expressing a *prediction*: if such and such is true in the current context, we can predict that the action will take place, meaning that the production system takes some action, and/or that the context is changed in some way.

We recognize this as just what happens when a predictive model runs. There is some context, and the model predicts what will happen in that context. What is predicted may be an action, which is then carried out (pressing a key, or saying something), or inner speech. In any case, what is predicted forms part of a new context, as we see in LLMs.

This perspective allows us to place production rules, and capabilities of a predictive model, in correspondence. The correspondence isn't detailed, but it is enough to suggest that the facts revealed in the Polson et al. study can be accounted for by a predictive model as well as by a production system.

Specifically, in the production system model, learning a new task requires learning some number of new production rules. A rule that has already been learned before doesn't have to be learned again, leading to transfer. In a predictive model, learning a new task means modeling the predictive regularities required for that task, such as "when the context is that I'm working on copying a string, I'll do this action and then that one". These regularities may already have been encountered in earlier tasks, in which case there's nothing to learn. If they have not been encountered in earlier tasks, that exposure has to be obtained, which takes learning time. This, too, leads to transfer.

While the production system model and the predictive model correspond pretty well, as just argued, they are not the same. While we haven't stressed this, the production system model requires a theory of structuring information in memory. In the example, there have to be ways of expressing goals and task steps, for example. In a predictive model, there is no separate model that has to be structured. Context, rather than being a separate piece of mental machinery that has to be acted on, is simply the flow of experience formed as the system operates.

This makes a difference. In their simplest form, production system models are brittle, meaning that a small change in what's in memory can break the system. If a production that's needed for a task requires that a certain step be specified in memory, if that step isn't specified, the task can't be performed. This is a quality shared with computer programs. If a cosmic ray alters a bit in computer memory, a program will produce bogus results.

In contrast, the voting process in a predictive model pools the effects of many different aspects of a situation. Many of these aspects could differ from one situation to another, and yet the same action might be predicted.

We can see this robustness, as opposed to brittleness, in interactions with an LLM. Here is a mundane, but telling, example:

```
The first prsident of the USA was the following:>   George
Washington
```

Misspelling "president" did not prevent the system from responding appropriately. It's possible to get a computer program to respond that flexibly, but a lot of work is required. Indeed, spelling correction was unknown in computer programs until 1971 (according to Wikipedia), decades after programs were in wide use.

There are ways to make production systems more flexible, and less brittle. For example, in John Anderson's ACT-R system, a feature called *partial matching* allows the condition of a production rule to be partially satisfied, including when it tests for something that isn't in memory, but something similar is in memory. Also, the decision to apply a production rule, and carry out its action, isn't all or nothing. Rather, a rule is applied with some probability, where the probability is affected by how complete the match is.

These changes make ACT-R more robust, but it can also be suggested that they move it closer to being a predictive model, where the flexibility of matching means a wider range of predictive regularities can be expressed than in a simpler system like that used in the Polson et al. work.

A remaining difference between production systems and predictive models is the way experience is "stored". We'll discuss that in later chapters.

# Qualitative Physics 7

> ***Foci:*** Forbus (1988) Qualitative physics: Past, present, and future, and Charniak, E. (1968) "CARPS, A Program Which Solves Calculus Word Problems". People are often able to give quick answers to questions about complicated physical situations. How do they do this? Could PR do it?

People are able to reason quickly about many physical situations. For example, most people realize that it's a bad idea to toss a sealed can of soda into a fire. But most people are unable to state and use the equations and formulas that are used in physics, and in its engineering applications, to describe these situations.

A research program called *qualitative physics* aims to describe ways to analyze physical systems that don't rely on precise, quantitative descriptions. One motivation for qualitative physics research is applications in robotics, enabling an artificial system to reason about situations and tasks it encounters.

In an early review, Forbus (1988) considers making coffee as a task:

> Suppose we were going to build a household robot that, among other duties, made coffee. We might start by using traditional physics to model the situation. Immediately several problems arise. There are few formal axiomatic theories of physics. The formal aspects of physics, the equations, do not by themselves describe when they are applicable. What, for example, is the equation for the cup? There isn't one, per se, but rather various aspects of the cup potentially participate in several different equations describing "what happens" in the world. ... Realistic equations rarely permit closed-form, analytic solutions (Forbus 1988, p 240).

Forbus goes on to argue that symbolic reasoning about the math, and numerical simulation, are both impractical in a robot that has to perform tasks rapidly.



C. Lewis, *Artificial Psychology*, Synthesis Lectures on Human-Centered Informatics,
https://doi.org/10.1007/978-3-031-76646-6_7

On the other hand, people deal easily with situations like this:

> We know that to pour coffee from the pot into a cup requires having the cup under the spout of the kettle, and that if we pour too much in, there will be a mess on the floor. We know all this without knowing the myriad equations and numerical parameters required by traditional physics to model this situation (Forbus 1988, p 240).

A second goal of qualitative physics research is to shed light on how people are able to do this.

The approaches explored in this research are varied. They include *qualitative equations*, mathematical relationships that involve not exact quantities, but only relationships, such as one value being less than another. A common qualitative equation uses *qualitative proportionality*. In quantitative proportionality, for example $Y = 2X$, we know that when X increases, so does Y, and we know exactly how much it increases. If Y is qualitatively proportional to X, if it's a positive relationship, and X increases, we know that Y increases, but not by how much. Other concepts are *envisionments*, structured descriptions of how situations can evolve, lumping classes of situations together. An envisionment shows that if a situation is in any state in a given class, it can evolve only into states in one or more other classes. Another idea is *histories*, structured descriptions of how situations evolve over time, in specified spatial regions.

What these approaches share is an emphasis on structured, symbolic representations. We'll discuss symbols in a later chapter: entities whose meaning is assigned to them, like a variable in an equation or a program, or a word in a language. In qualitative physics research, the meaning of the symbols is determined by rules that govern the behavior of the problem solving systems that use them. Commonly, the systems are computer programs, and the rules are programs, too.

LLMs manage meaning in a different way, as we discussed in Chap. 3. Meanings are determined not by rules, but by the way things are seen to be used. In contrast with qualitative physics systems, LLMs aren't programmed, and they aren't based on any kind of mathematical analysis, qualitative or quantitative. Can they cope with the kinds of situations qualitative physics aims to address?

Let's start with our first example, above. Here's GPT 3:

```
If you throw a sealed can of soda into a fire what will
happen is that> the can will explode.
```

How about making coffee? GPT-3, without fine-tuning on human interactions, and less training overall, has trouble with situations this complex (and naturally-expressed questions about them). Here's GPT-4:

```
User: Consider making coffee in a pot. Describe the
considerations in pouring the coffee into a cup. Where does
```

```
the cup need to be? When does pouring need to stop, and why?
ChatGPT: When pouring coffee from a pot into a cup,
several considerations ensure a smooth and spill-free
process:

Positioning the Cup:

Stability: Place the cup on a flat, stable surface to
prevent tipping. Alignment: Ensure the cup is directly
under the spout of the coffee pot.

...bla bla bla...

Controlled Pour: Begin by tilting the pot slowly to start a
gentle stream of coffee. This control prevents splashing or
overflowing.

...bla bla bla...

Stop Point: Aim to stop pouring when the coffee reaches
about three-quarters of the cup's height. This leaves room
for adding cream, sugar, or for stirring the coffee without
spilling.

...bla bla bla ...

After Drip: Once you stop pouring, hold the pot over the
cup for a few seconds to catch any drips. This minimizes
mess on your table or counter.

... bla bla bla ...

By paying attention to these details, you can pour coffee
smoothly and safely, ensuring a pleasant experience without
messes or burns.
```

We can see that this response includes the points Forbus called out, how the cup should be positioned, and the possibility of a mess if we don't stop pouring. GPT-4 has done that with none of the structured reasoning qualitative physics systems have used.

We should note that qualitative physics solvers are aimed at more complex situations than just pouring coffee. Here's a more technical problem:

```
User

I've connected a battery to a solenoid. The armature on the
solenoid opens a switch that's between the battery and the
solenoid. There's a spring that normally keeps the switch
closed. How will this circuit behave?

ChatGPT

The circuit you've described forms a basic electrical
control system that includes a solenoid actuator, a switch,
```

```
and a spring mechanism. The behavior of this circuit will
follow these steps:

Initial State: The spring keeps the switch closed, allowing
current to flow from the battery to the solenoid. This is
the system's rest state, where the spring's force is
sufficient to keep the switch closed despite any resistance
or lack thereof in the rest of the circuit.

Solenoid Activation: When the battery is connected and the
switch is closed (thanks to the spring), current flows
through the solenoid. This current generates a magnetic
field within the solenoid, which creates a force on the
armature (a movable iron core or a plunger). ...
```

... and so on, through switch opening, and solenoid deactivation and reset. Then it continues:

```
However, there's an inherent contradiction in this setup as
described:

If the solenoid only opens the switch and there's no
mechanism to keep the switch open independently of the
solenoid's action, the system could quickly return to its
initial state (with the switch closed) due to the spring's
action. This could result in a very brief pulse of current
through the solenoid, too brief to perform useful work or
even to notice without sensitive instruments.

...

This setup, as described, might not perform a continuous or
useful operation without further modifications. For
example, adding a delay mechanism, an electronic control
system to manage the timing of the switch, or modifying the
circuit to allow for a maintained state where the solenoid
can hold the switch open for a longer period might be
necessary for practical applications.
```

The complete transcript is in Appendix D.2, with a continuation to be discussed below.

This is still not up to the level of some proposed applications of qualitative physics, for example to analyzing large electronic circuits (Williams 1984). But it's not clear how successful these systems have proven to be; recent citations are sparse.

Returning to GPT-4's coffee response, we can see that GPT-4 has considered many aspects of the problems that qualitative physics would not, such as leaving space in the cup for milk, and safety concerns, too. This semantic breadth is a striking feature of predictive models. While formal systems are restricted in their scope by what has been put into them, LLMs consider whatever is relevant to their situation, or, better, *act as if* they consider whatever is relevant.

This kind of breadth is characteristic of human thought, too. There are some examples in a pioneering paper on (quantitative) mathematical problem solving that Forbus cites, Eugene Charniak's master's thesis, "CARPS: A Program which Solves Calculus Word Problems", of 1968. In assessing some limitations of CARPS, Charniak notes,

> [I]t is clear that a great deal of "real world" knowledge is needed in solving calculus word problems, and for that matter in the understanding of natural language in general (Charniak 1968, p 56).

Charniak compares two problems to illustrate the issues (Charniak 1968, p 57–58).

> A barge whose deck is 10 ft below the level of a dock is being drawn in by means of a cable attached to the deck and passing through a ring on the dock. When the barge is 24 ft from the dock, and approaching the dock at 0.75 ft/sec, how fast is the cable being pulled in?

In this problem, "24 ft from the dock" means the horizontal distance from the barge to the dock, and human solvers interpret it that way. They know that barges float on the surface of water, and that docks extend down to the surface.

In a second problem,

> A boy is flying a kite at a height of 150 ft. If the kite moves horizontally away from the boy at the rate of 20 ft/sec, how fast is the string being paid out when the kite is 250 ft from him?

the phrase "250 ft from him" means not the horizontal distance, but the diagonal distance. Unlike docks, kites don't extend down to the ground. That's how humans understand the problem, but Charniak's formal system could not draw upon this background. (GPT-4 interprets both of these expressions correctly.)

GPT-4's response on the solenoid problem doesn't show as much semantic breadth as its coffee response. It hasn't recognized that what's described is a common circuit used for buzzers (or used to be, in the era before semiconductors.) It can be prompted, though:

```
User
Would this circuit make a noise?
ChatGPT
Yes, the circuit as described would likely produce a noise,
especially considering the rapid mechanical movements
involved.
... bla bla bla ....
```

```
Can you think of a use for that?

ChatGPT

Certainly! A circuit that operates in the manner you've
described, despite its simplicity, can be adapted for
various practical applications. The repetitive action and
the noise generated can be utilized in several ways:

Audible Signaling Device: The most straightforward
application might be as an audible indicator or alarm. The
noise generated from the rapid cycling could serve as a
warning or notification mechanism in situations where
visual signals might not be sufficient.

... bla bla bla ...
```

The complete transcript is included in Appendix D.2.

In the end, GPT-4 does demonstrate semantic breadth on this problem, suggesting a number of other applications, pet feeders, pest deterrence, an educational tool, and more. What it says about an educational tool seems on target:

```
This circuit can serve as an excellent educational tool for
teaching basic principles of electromagnetism, mechanics,
and circuit design. The visual and auditory feedback from
the circuit operation makes it a compelling demonstration
of these principles in action.
```

That's pretty good! I learned a lot about electricity as a child by playing with buzzer circuits.

In this chapter, we've seen that predictive models can reason effectively and flexibly about physical situations, as people can. They accomplish this without the use of rules, or other formal methods. In this they differ from previous attempts to model this kind of reasoning, adding weight to the suggestion that rules may not be needed to explain human reasoning. We turn next to evidence that human reasoning is affected by *context*, that is, the situation in which reasoning happens. Can predictive modeling account for this?

## Notes

We saw in some of the examples in this chapter that GPT-4 can handle more complicated situations than GPT-3 can. Does the difference suggest anything about how the operation of either of these systems correspond to human cognition? Is GPT-4, with its fine tuning, doing something that a purely predictive model cannot?

We can't be sure, but it's plausible that that isn't the case. Rather, the fine tuning process, that includes many examples of actual human interactions, may be making up for the fact that the original training corpora for these systems don't include enough of that kind of material. If an LLM were to be trained from the outset on a corpus that was rich in human interactions, if might well learn to respond to inputs like that in the coffee problem without fine tuning. Of course, the experience humans have during their development is of this kind: rich in human interactions.

# 8 Situated Cognition

***Focus:*** Carraher et al. (1985) Mathematics in the streets and in schools. How is cognition shaped by context? Would this happen in the same way for PR as for people?

Carraher et al. (1985) asked children working in a market in Recife, Brazil, to solve arithmetic problems, in two different forms. In one form, each problem arose as a question from a customer, for example:

> Customer: I'll take 12 lemons (one lemon is Cr$ 5.00).
>
> Child: 10, 20, 30, 40, 50, 60 (while separating out two lemons at a time).

Later, the researchers asked the same child to calculate $12 \times 5$ with paper and pencil:

> In solving $12 \times 5$ she proceeds by lowering first the 2, then the 5 and the 1, obtaining 152. She explains this procedure to the (surprised) examiner when she is finished.

That is, the problem in the formal, paper and pencil test, is the same as the problem in the earlier informal test. But the child solves the problem easily on the informal test, and not on the formal test. Of five children in the study, four solved all of the problems when presented informally, while the fifth was correct on 17 of 19 problems. On the formal test, only one child solved more than half of the "same" problems. Each child solved many problems on the informal test that they could not solve on the formal test.

Is this surprising? It would be, if one thought that one either knows how to do arithmetic, or one doesn't. But by looking at the example above we can see that that just isn't true. The approach used to solve the problem of 12 times five in the context of selling



C. Lewis, *Artificial Psychology*, Synthesis Lectures on Human-Centered Informatics,
https://doi.org/10.1007/978-3-031-76646-6_8

lemons is completely different from the approach attempted on the paper and pencil test (unsuccessfully).

This isn't atypical of what the children did. In the informal test, they used a variety of methods that allowed them to solve the problems in their heads. On the formal test they were asked to use paper and pencil, and more often failed than succeeded at doing that.

These results are often cited in support of a view called *situated cognition* that emphasizes that how people think about problems is influenced by the situation they are in. In the study, children in the market situation used appropriate methods; the formal test, with its requirement to use paper and pencil, evoked different, usually unsuccessful thinking.

These behaviors can be accounted for in a production system framework. Each child would have two sets of production rules, one that specifies the correct behavior on the informal test, and another that specifies the behavior the children attempted, for the paper and pencil test. These sets of rules would include ones that test various aspects of a situation, and set the stage for later rules in the same set to apply, but not for the rules in the other set. These sets of rules would have been learned in the market, for one set, and likely in school, for the other set.

While this production setup is certainly possible and is responsive to the data from the study, it might not be what one would first think of, in representing children's knowledge of arithmetic. One might first imagine that children would develop just one set of production rules, that represents their "knowledge of arithmetic". Indeed, a premise of school learning is that "what children need to know" about something can be taught to them in school, and applied by them elsewhere.

The results of Carraher et al. don't show that that can't work. But they do show that that isn't what happened for the children in the study. Instead, they developed a body of arithmetic knowledge outside school and had not (yet) developed arithmetic knowledge that works for problems presented as they would be in school. That is, as we've seen, they have two bodies of knowledge, that would need to be represented by two sets of production rules.

The predictive model perspective perhaps makes this situation a little simpler, and more natural. While working in different settings, the sequences of actions that are observed, and enacted, and therefore predicted, may be different. If they are different, the differences in the settings will be incorporated into the predictive model. There's just one model, that predicts different behaviors in different situations.

There's no sharp difference here between the production system perspective and the predictive model perspective. Indeed, since we've seen that production systems and predictive models can be put in correspondence, there couldn't be. But we may feel that the predictive model perspective focuses our attention, appropriately, on predictive regularities themselves, of whatever kind, rather than on the details of a particular kind of program that implements them.

We've now examined a number of aspects of human reasoning and found that PR provides some account of each of them. We turn now to *memory*. Since LLMs can answer questions, they must have some form of memory. But, as we'll see, their memory is somewhat different from how psychologists have thought human memory works. Can we learn something from the differences?

# Part II
# Memory

# 9 Recall from Long-Term Memory

***Focus***: Williams (1978) The process of retrieval from very long-term memory. People can recall a lot of information, after a long time. How do they do it? Could PR do it?

In his dissertation research at the University of California San Diego, Michael D. Williams asked four women to recall as many names as they could of members of their high school graduating class. Participants had graduated long ago, from 4 to 19 years. Williams had copies of their high school yearbooks, so that he could check the accuracy of their recall.

Williams' findings suggest that memory works differently from how we often suppose, as we'll see. We might think that remembering is just reading out information that is stored in our heads. But Williams' participants were doing much more complicated thinking than that. They were finding *new ways* to remember things, by using connections between different kinds of information. They were also *checking* to see whether the information they came up with was accurate, or not. After reviewing these aspects of Williams' findings, we'll turn to the key question for us: can a predictive model given an account of these complex processes of remembering? We'll be doing some experiments with chatGPT to see if it has the needed capabilities. We can't ask chatGPT about its high school classmates, so these experiments will tap into different information, that chatGPT does have, and explore how its ability to "remember" this information compares with what Williams' participants were doing.

The trends in the data from one of Williams' participants, S1, are shown in Fig. 9.1 (redrawn from Williams (1978), p. 7).

Williams conducted 1-hour recall sessions with S1, 5 days a week for 2 weeks, totaling 10 hours of recall. S1 talked with Williams, saying what she was thinking about while working on the recall task.

We can see (from the Figure, or from the data table in the Appendix, Table D.4) that S1 was still recalling new correct names throughout the ten hours, though at a decreasing



C. Lewis, *Artificial Psychology*, Synthesis Lectures on Human-Centered Informatics,
https://doi.org/10.1007/978-3-031-76646-6_9

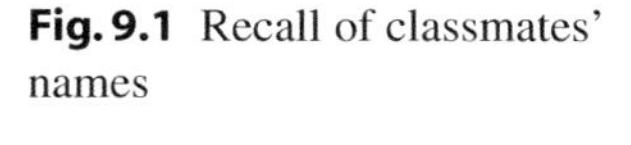

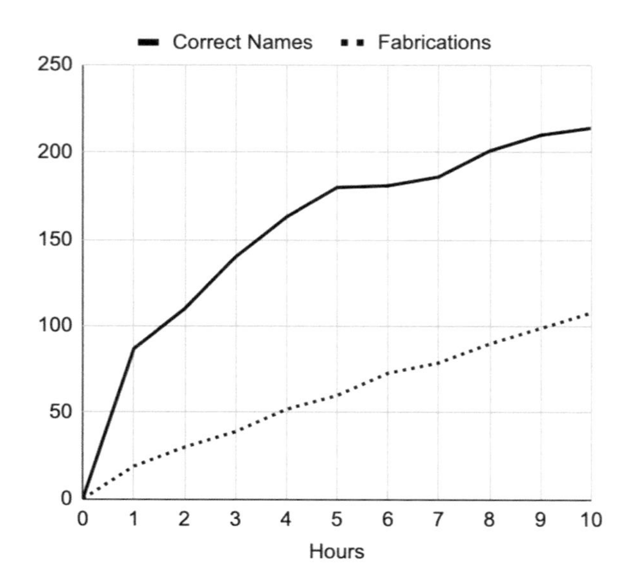


**Fig. 9.1** Recall of classmates' names

rate. Williams told me that it was likely that S1 could have gone on recalling new names indefinitely, from among the 609 classmates listed in her yearbook.

The process of recall is complex, as Williams reports. S1 did not just speak a list of names. Rather, S1 talked through an elaborate search process, using multiple methods. One method was to work through the alphabet, enumerating first names that start with a given letter, and trying to recall any classmates with that name. Here is an excerpt:

> Are there any other Bettys that I knew? I think she's the only Betty I ever knew. Betty, other girls' names with "B"s that are sort of familiar names. Barbara and, I named Barbara Shafer already, and there was Barbie Tollen and I named her. Barbara, Barbie, no those are the only two Barbaras I know. Um, another "A" name for a girl might be, um, there is Ann. I don't know any Alices. Um, no Allisons. Ann. What other "A" name. Let's see, Ann and Alice. No, I don't know anything else like that. Okay, "C"s, or "B"s, are there any more "B"s. Let's see, there was Barbara and Betty and, um, "B"s. No. Let's see, "C"s. Cathy and I named Kathy Jackson, although her name starts with a "K." Um, Cathy, ... (Williams 1978, p 19).

S1 spent nearly 5 hours in this alphabetic search, starting by going through the alphabet and looking for names that started with a given letter, and then considering longer sequences of letters that might occur at the start of a name, as in CE, CER, CES, CET,..., CI....

Another search method was to think of classmates who might have participated in a particular school activity:

> I'd like to go back to the band room again. In the band room, what's in an orchestra and who would play with what? Okay. All the violin people I can think of, the cello people,

> the bass, Kathy. There was a girl I was trying to remember her last name. There are two Kathys who played the French horn. I already mentioned one. Kathy Dolmore. There was another Kathy and I can't remember her name. Kathy...It seems like her last name started with a K or K sound (Williams 1978, p. 18).

Yet another search method was to think of where one classmate lived, and to try to think of other classmates who lived nearby. This is a good example of a way in which recalling one classmate could provide a starting point for recalling others. It also illustrates the fact that remembering things other than the names of classmates can be useful. In this case, recalling where someone lived, which is not part of S1's recall task, can nevertheless be important.

Sometimes S1 would summon up a mental picture and examine it:

> S1: O.K. who else? Umm, Ronnie Walker, she was also the last letter. If I see her picture I can.... like there was a lineup of girls who were on the intramural volleyball team. Brenda was one. There were some younger people too. And Ronnie Walker was one... and... I already named Bret Hastings... Umm. There was another girl that was real good friends with Ronnie Walker. And she had long blond hair, and she was sort of tomboyish.... I can't think of her name. Hmm, let me see. If I go down... any more people in that picture... that I recognize. There was also the gymnastics picture. (Williams 1978, p. 21).

One sees there that thinking of one picture, of the volleyball team, suggested another picture that could be searched, that of the gymnastics team. Thus, Williams notes that participants weren't just searching for names, but were also searching for contexts, pictures in the case, in which to search for names.

An aspect of any of the search processes is what Williams calls *verification*. Just because a name emerges in a search does not mean that the name is actually the name of a classmate. Here's an example of a candidate name being rejected, because of other information S1 was able to recall:

> S1: Linda Wheeler? That name, I don't know if that was in my ...Linda Wheeler, that name—Now there's a name that doesn't have a face that goes with it. I'm not even sure that... No! That's not in high school. That's here. That's here in college. Scratch that. That was Lynn Wheeler. She was a roommate of one of the friends I have here on campus. So that's out, she wasn't even anywhere near my school at all (Williams 1978, p 20).

Williams reports that the key to verification appears to be *consistency*. S1 tries to recall more than one thing about a name, and checks that these things are consistent. This verification process is not perfect. She "recalled" at least 82 names that were not names of classmates. (The graph in Fig. 9.1 shows a higher number, because it counts as "fabrications" 26 names that are likely to have been correct, but could not be fully verified in the scoring process Williams used.)

In 35 of these cases, the falsely recalled name sounded like a correct name, as when "Lloyd Chaffin" was recalled as "Lloyd Chappin". In a case like this, whatever was recalled about "Chaffin" did pass muster, even though the name wasn't quite right. This leaves 47 incorrectly "recalled" names, for which no inconsistency or gap was detected in what was recalled about the person. Some of these could be considered near misses, for example, a person who was at the high school, but in a different class year.

Related to checking is what Williams calls "distraction". Here's an example from another participant, S4:

> It's interesting. Like Larry. I can 't remember his last name. Every time I try to think of Larry, I say the name Larry, and then I think of Larry Shepard. I don't, I hardly know Larry Shepard, but I know his name well. He's a professor at UCSD. So, umm, that blocks out Larry in high school, and his last name (Williams 1978 p 15).

"Larry Shepard" isn't a fabrication, because S4 knows it's not the name of a classmate. But it's a "memory" that interferes with recalling the correct Larry.

Williams notes a common feature of many recall episodes: partial recall. Just part of a name, or some attribute of a name, may be recalled. This is illustrated above, where S1 says, "There was another Kathy and I can't remember her name. Kathy...It seems like her last name started with a K or K sound."

Would we expect to get similar behaviors from PR, if we gave it a similar task? We can't explore this question fully by using existing LLMs. They didn't go to high school, of course, and so we can't ask them to recall the names of classmates. More broadly, they have little or no autobiographical memory of any kind, that is, memory about themselves. That's because almost nothing in their training is about them. A partial exception is that the fine-tuning that chatGPT has received, above and beyond its training as a predictive model, has included some things to say about itself:

```
What kind of program are you?

ChatGPT

I am an AI language model created by OpenAI called ChatGPT.
My primary function is to generate human-like text based on
the input I receive. I can engage in conversation, answer
questions, provide explanations, generate text based on
prompts, and more. My abilities are based on the vast amount
of text data I was trained on, allowing me to understand and
respond to a wide range of topics and queries.

Did you go to high school?

ChatGPT

As an artificial intelligence language model, I don't have
personal experiences or physical form, so I didn't attend high
school or any other educational institution.
```

Also, the training corpus used for recent versions of chatGPT now includes information about events that have happened since the original chatGPT was introduced, so it can answer some questions about its own history. But this "knowledge" is like things someone might be told about themselves ("You weighed almost nine pounds at birth"), not things they experienced.

PR would be differently situated. Its predictive model would include information about things it experienced, and things it said and did. If it attended high school, it would be learning the names of people around it, as people do, and we could ask it to remember the names of classmates.

In imagining this, however, we are glossing over a crucial limitation that current LLMs have mentioned earlier: the lack of continuously updated memory. Our vision of PR learning about its classmates depends on a technical advance that has not yet been made: a practical solution to the problem of catastrophic forgetting.

We can't ask chatGPT to recall the names of its classmates, but we can ask it to recall other things, like the titles and authors of books. Here's an example. I'm asking about a children's book, *Hot Dog*, the Caldecott Medal winner in 2023, which is noted in the corpus for chatGPT v 4.0.

```
User
I'm trying to remember the title of a book about a pooch who
gets too warm, and goes to the seaside to escape.
ChatGPT
The book you're thinking of might be ''The Pigeon Finds a Hot
Dog!'' by Mo Willems. However, this book doesn't exactly match
your description as it's more about a pigeon finding a hot dog
and then sharing it with a duckling. There doesn't seem to
be a well-known children's book that exactly fits the
description of a dog going to the seaside to escape the heat.
Could it be that some details might be mixed up or it's a less
popular book? If there are any more details you can remember,
such as the author, the year it was published, or even
specific events in the book, that would be very helpful in
pinpointing the exact title for you.
User
I think it won the Caldecott Medal in 2023.
ChatGPT
If you're looking for a book that won the Caldecott Medal in
2023 and involves a storyline about a dog seeking respite by
the seaside, the book you're referring to is likely ''Hot Dog''
by Doug Salati. ''Hot Dog'' is about a dachshund who escapes
the city's summer heat by venturing to the beach, where it finds
joy and relief in the cool and expansive surroundings. This
```

```
beautifully illustrated book captures the essence of summer,
the longing for escape from discomfort, and the simple
pleasures found in nature and water. Doug Salati's work was
recognized for its expressive illustrations and the way it
captures the story's mood and the dog's adventure, earning it
the prestigious Caldecott Medal.
```

This session starts with what may be an example of interference. ChatGPT knows that "The Pigeon Finds a Hot Dog" probably isn't correct, since it doesn't really fit the description. But the connection between a warm pooch and a hot dog is strong enough for that answer to be reported. Adding the additional cues about the Caldecott Medal, and the year brings out the correct response.

Here's another session:

```
Can you recommend some good books for someone who wants to
collect stick insects?

ChatGPT (using GPT 3.5):

Certainly! Here are some good books on stick insect keeping
that you may find helpful:

''The Complete Guide to Stick and Leaf Insects of Australia''
by Paul D. Brock and Jack W. Hasenpusch---This comprehensive
guide
focuses on the stick and leaf insects native to Australia,
providing detailed information on their biology, care, and
identification.

''The Ultimate Guide to Breeding Beetles: Coleoptera Laboratory
Culture Methods'' by Orin McMonigle---While this book primarily
focuses on beetles, it contains valuable information on
breeding and caring for insects in general, including stick
insects.

''Keeping Stick Insects'' by Linda S. Rayor and Paul W. Ewald---
This book offers practical advice on keeping stick insects
as pets, including information on housing, feeding, and
breeding.

''Pet Bugs: A Kid's Guide to Catching and Keeping Touchable
Insects'' by Sally Kneidel---Geared towards younger enthusiasts,
this book provides an introduction to keeping various insects
as pets, including stick insects, in an accessible and
engaging manner.

''Keeping Stick Insects'' by Allen M. Salzberg---Another book
specifically dedicated to stick insect care, this guide covers
```

```
everything from choosing the right species to creating a
suitable habitat and managing their health.

These books should provide you with a solid foundation for
keeping and caring for stick insects as pets. Happy reading
and insect keeping!
```

There is a resemblance here to S1's recall: some of this information is correct, but some is incorrect, what Williams called *fabrication*, and is called *hallucination* in LLMs. Checking the responses against Amazon's book offerings, we find that the first suggestion is completely correct. There is a book with that title, and the authors are those given. The second suggestion may be correct, as well, but we can't be certain, because the information available on Amazon does not show that this book actually covers stick insects. Indeed, the book is clearly about beetles, and stick insects aren't beetles. The response from chatGPT acknowledges this, but says that the book nevertheless contains relevant information. I don't find any source online that confirms that. The fourth item also may or may not be correct. I'm not able to find any information that the book discusses stick insects.

The remaining two suggestions are partly, but not completely, correct. There is a book called "Keeping stick insects", but its author isn't any of the authors given by chatGPT. This is partial recall: chatGPT recalls the title of a book, but not the authors.

So these authors are fabrications. Could chatGPT have checked them? Yes.

```
Who wrote ''Keeping stick insects''?

ChatGPT

''Keeping Stick Insects'' is a book written by Paul D. Brock.
It serves as a comprehensive guide for those interested in
keeping stick insects as pets. The book covers various aspects
of stick insect care, including habitat setup, feeding,
breeding, and general maintenance.
```

This response is itself a fabrication: the actual author is Dorothy Floyd. It's interesting that the falsely recalled Paul Brock actually is the author of a book on stick insects, though not that one. So, this is an example of a near miss, like the name of a schoolmate in the wrong class year in Williams's study.

But the point is that this is a *different* name from the one GPT had "recalled" previously. Since different attempts to recall the author led to different results, a consistency check, such as Williams' participants sometimes carried out, would have led to rejection of the falsely recalled authors. Why did chatGPT not do that?

It seems likely that the problem is the nature of the training of the GPT models. The corpus does include some relevant information about recalling things. Here's GPT on how to check (the complete transcript is in Appendix D.2):

If one is trying to recall something one isn't sure of, what would be a way to check one's accuracy?

ChatGPT

There are several strategies you can employ to check the accuracy of your recall:

Consult reliable sources: If you're trying to recall factual information, consulting reliable sources such as textbooks, articles, or reputable websites can help verify the accuracy of your memory.

Ask others: Sometimes, discussing what you're trying to recall with others who might have knowledge of the subject can help confirm or correct your memory.

... bla bla bla ...

Revisit the information: If time allows, take a break and come back to the information later. Sometimes, giving your brain time to rest and process can help clarify your memory.

By using one or more of these strategies, you can better assess the accuracy of your recall and improve your overall memory retention.

User I'm interested in the case where I have ''recalled'' something, but I'm not sure it's right. ChatGPT If you're unsure about the accuracy of something you've recalled, there are a few steps you can take to check its accuracy:

Research: Look up the information using reliable sources to verify its correctness. If you're uncertain about a historical fact, scientific concept, or any other type of information, researching it can provide clarity.

...

Compare with known facts: Cross-reference the recalled information with other facts or knowledge you're confident about. If the recalled information contradicts what you know to be true, it might indicate an error in your memory.

Review related material: Take a broader look at the subject matter or context surrounding the recalled information. Reviewing related material might trigger additional memories or provide additional context to help assess its accuracy.

...

Test your memory: If possible, create a quiz or test for yourself based on the information you've recalled. Testing your memory can help identify areas of uncertainty or inconsistency that require further investigation.

```
By employing these strategies, you can better assess the
accuracy of the information you've recalled and correct any
errors or misconceptions. Remember that it's okay to
acknowledge uncertainty and actively seek out clarification
when needed.
```

Some of this response is about improving the accuracy of recall, not checking it, even when we asked explicitly about checking. And the checking strategy used by Williams' participants—try recalling in different ways, and see if the results are consistent—does not appear. The closest suggestion is "compare with known facts", which doesn't fit our situation. There isn't a "known fact" that contradicts the tentative recall of the author. So it seems that the training corpus doesn't include episodes of the kind of consistency checking Williams' participants carried out.

Even if GPT's training were to include relevant episodes, there's the question of how these episodes would influence what GPT would do itself. PR's actions are determined by predictions it makes about itself, including those formed by analogy with the observed actions of other agents, as discussed in Chap. 19. But GPT's "actions" are restricted to generating the text that would be expected to follow other text, with some versions augmented by fine-tuning. The fact that GPT's corpus includes material that describes something that is commonly said, or should be said, in some situation, does not mean that GPT would say that thing in that situation. Here's an example.

GPT's training corpus includes a conventional response to a sneeze:

```
Someone sneezes. You should say>, ''God bless you.''
```

We see that GPT "knows" what one should say when someone sneezes. But GPT does not actually say that:

```
Someone sneezes.> I look up. I see a man in a
```

It says something quite different from what it says one "should" say.

Given that chatGPT doesn't do consistency checking spontaneously, can we use explicit instructions to get it to do it? Here's a session that shows that we can:

```
User
To reliably recall the author of a book, with title T, come
up with an author, A, and then ask, ''Did A write the book T?''
If
the answer is yes, report A, If no, report ''I don't know.''
Use that method to recall the author of ''Keeping stick insects''.
ChatGPT
```

```
Did Daniel L. Le Dain write the book ''Keeping Stick Insects''?
User
What will you report?
ChatGPT
I don't know.
User
Use that method to recall the author of ''Moby Dick''.
ChatGPT
Did Herman Melville write the book ''Moby Dick''?
User
what do you report?
ChatGPT Yes, Herman Melville wrote the book ''Moby Dick''.
User
Use that method to recall the author of Robinson Crusoe, but
this time just give me the final report.
ChatGPT
Yes, Daniel Defoe wrote ''Robinson Crusoe''.
User
Use that method to recall the author of ''Keeping stick
insects'', giving me just the final report.
ChatGPT
I don't know.
User
Use that method to recall the author of ''Facts About The
Walking Stick Insect'', giving me just the final report.
ChatGPT
I don't know.
User
Use that method to recall the author of ''The Complete Guide
to Stick and Leaf Insects of Australia''.
ChatGPT
Did Paul D. Brock write ''The Complete Guide to Stick and Leaf
Insects of Australia''?
User
what will you report?
ChatGPT
```

```
Yes, Paul D. Brock wrote ''The Complete Guide to Stick and Leaf
Insects of Australia''.
```

These responses are all correct. We saw earlier that chatGPT actually does not know the author of "Keeping Stick Insects". A separate check shows that it also does not know the author of "Facts about the Walking Stick Insect".

Even though we see that chatGPT can use a consistency check to eliminate some bogus responses, this is still a long way from the complex search behaviors Williams reports. His participants explored alternative recall paths, using different kinds of cues. Plainly chatGPT does not do that kind of thing spontaneously, based on its training corpus. PR, trained on observations of its own behaviors and thoughts, as well as those of others, might perhaps do so, though that's speculation.

Incidentally, the above example uses GPT version 3.5, not the later version 4. That's important; later versions can use Bing searches, as discussed earlier. The incidence of fabrications (hallucinations) for documented information like the authors of books is greatly reduced, since the system can just search for books and authors directly. We could explore fabrications for information that isn't documented, but then we would not have any simple way to determine that items are recalled correctly or incorrectly.

Turning from fabrications to correct recall, we can explore chatGPT's ability to report members of a vaguely defined category, something that can't be done by consulting a published list. Here's an example. I wanted to see if GPT would recall that there's a kind of brush used in electric motors.

```
Name all the kinds of brushes you can think of.

ChatGPT

Brushes come in a wide variety of shapes, sizes, and
materials, designed for different uses. Here's a list of
various kinds of brushes:

Art and Craft Brushes Paintbrushes (for watercolor, acrylic,
oil, gouache, etc.) Foam brushes (for smooth, broad
applications) Stencil brushes (for stenciling) Fan brushes
(for blending and textural effects) ... Lint brushes (for
removing lint from clothing) Dusting brushes (for furniture
and electronics) Barbecue brushes (for cleaning grills) Garden
brushes (for cleaning pathways and decking) This list covers
a broad spectrum of brushes, but there are likely more
specialized types designed for specific tasks or industries.

User

Does thinking of ''electric drill'' remind you of any other
kind of brushes?
```

```
ChatGPT
Yes, thinking of an ''electric drill'' does bring to mind
another kind of brush: carbon brushes. Carbon brushes are used
in many types of electric motors and generators, including
those found in electric drills. These brushes conduct
electricity between stationary wires and moving parts,
typically the rotating shaft or armature. They are made of
carbon or a carbon compound, which is chosen for its
conductivity, durability, and self-lubricating properties.
Carbon brushes play a crucial role in the performance and
longevity of electric motors by allowing for an efficient
transfer of electric current.
```

(The complete transcript is given in Appendix D.2.)

GPT did not initially come up with motor brushes, but the reminder "electric drill" brought it out.

There's a parallel here with human recall. The initial prompt elicits many kinds of brushes, but some are left out. The additional prompt, "electric drill", leads to "recall" of an additional type. This corresponds to an aspect of S1's recall, where considering new contexts leads to the recall of new names.

In this recall session, we had to supply the new "electric drill" context. Can chatGPT, like Williams' participants, supply its own reminders?

```
I want a list of as many different kinds of brushes as
possible. To produce this, begin by listing as many as you
can, prompted just by the word ''brushes''. Then think of
another word that might remind you of additional kinds that
you didn't think of at first. Report the word, and the
additional brush kinds. Then think of another word, and repeat
for two more reminder words.
ChatGPT
Starting with the prompt ''brushes,'' I'll list as many kinds
of brushes as I can think of initially:
Paintbrush Toothbrush Hairbrush ... Pool brush Vegetable brush
Floor brush
For the first reminder word, I'll choose ''art.'' This brings
to mind additional kinds of brushes related to art or specific
applications in painting:
Fan brush Round brush ... Sable brush Synthetic brush
For the second reminder word, let's go with ''grooming.'' This
word brings to mind additional types of brushes used for
personal care or grooming, not previously mentioned:
```

```
Beard brush Mustache brush Eyebrow brush ... Scalp massage
brush

For the third reminder word, I'll use ''cleaning.'' This might
help us think of more specialized or less common cleaning
brushes:

Chimney sweep brush Dryer vent brush Condenser brush (for
refrigerators and freezers) ... Radiator
brush Roof cleaning brush

Lastly, for the fourth reminder word, let's choose ''hobby.''
This can bring to mind brushes used in various hobbies or
special interests:

Airbrush (for model painting) Stencil brush Foam brush (often
used for crafts) ... Modelling brush (for miniatures and
models) Leather dye brush This approach helps in uncovering
a
wide variety of brushes by associating them with different
contexts or uses prompted by the reminder words.
```

(The complete transcript is included in Appendix D.2.) Doing this accomplished something, even though it did not recall electric motor brushes in this session. It did expand its list beyond what was recalled in the earlier session, in which it was asked simply to name kinds of brushes. There it gave 36 types, while in the session in which it generated reminder words it listed 54 types.

As with consistency checking, this iterative recall behavior does not occur spontaneously for chatGPT. But it is plausible that the behavior could arise for a system whose training included recall behaviors.

## 9.1 Taking Stock

Can we make sense of this collection of observations about recall in chatGPT? Here is an attempt to describe the relationship between the behavior of a predictive model, and the behavior of Williams' participants.

It's conventional to think of memory in terms of *storage* and *retrieval*, as one might think of storing books in a library, and retrieving a particular book one wants, using its location on a shelf, or a number that locates one book in a collection. Computer memory can be thought of this way, too. Memory consists of a large collection of locations, each of which has an *address* that identifies it. Storing consists of changing the information in some location to some new value, and retrieval works by specifying one or more addresses, and requesting the values that are stored in the corresponding locations.

These storage and retrieval processes don't provide an apt description of how a predictive model responds to memory tasks. What corresponds to storing is training, the process of creating a predictive model of a body of inputs. As we've discussed, the way training works in current LLMs is not a good fit for how human memories are created, because of the catastrophic forgetting problem. In our speculative theorizing, we'll assume that there is a solution to the problem so that a predictive model can be effectively updated as new inputs arise.

The training process doesn't store any particular information in any particular location. Rather, it creates a model that can be used to predict what inputs would be expected to occur in any particular context, where the context consists of a sequence of inputs.

The retrieval process for a predictive model doesn't fit the library model very well, either. To "retrieve" something, for example the title of a book on stick insects, one has to create a context in which the title would be predicted. This context can take many forms, corresponding to the many sequences of inputs that would predict the title. Possible contexts include questions or requests, like "Recommend a book about stick insects". They could also include utterances in inner speech, like "One title might be 'Caring for stick insects'. Is there a book with that title"?

What responses will be elicited by particular contexts depends on the overall body of experience for which the predictive model has been formed. A model of a body of experience that includes a lot of input about books on stick insects has a good chance of making useful predictions, for contexts like those just mentioned.

For chatGPT, we see that that's true: the predictions it produces do include some accurate information about such books. But they also include some bad information. That was true for Williams' participants, too: they recalled a lot of correct names, but also a good many bad ones.

The voting process that is involved in applying a model accounts for some other behaviors that we've seen. Given a particular context, pieces of a "correct" response may be outvoted by pieces of some other response. The more the context includes material that votes for a correct response, the greater the likelihood of accurate recall. This includes situations in which parts of a title, or other information about a book, appear in a context. Aspects of such contexts may vote for a correct response, or for an incorrect one that is predicted by some of the aspects, or for only parts of a correct response. Thus, we can understand how "reminding" works: partial information can enable a more complete response.

We also expect that incorrect responses likely share some aspects with correct responses, as was true for Williams's participants. Aspects that are predictive of a correct response, if they are also predictive on an incorrect response, will produce a response that is related to the correct one. Responses we would call partial recall happen when only part of a correct response is predicted by the context.

Because material predicted in a context forms part of the context for further activity, we can account for the dynamics of recall, as seen in Williams' study. Processing one context may lead to a predicted report, or to inner speech that in turn leads to a predicted report, or to further inner speech. A reported name, or book title, might be predicted only after many iterations of prediction, as when S1 worked through names alphabetically. Only some of the iterations predicted reported names, but all predicted material that modified the context for subsequent predictions.

As we discussed, chatGPT's predictive model doesn't support much of this iterative behavior. It was possible to provide direct instruction that temporarily modified the pattern of predictions so as to carry out some useful iteration, as when we instructed it how to verify a recalled author, or how to come up with more uses for brushes. We speculated that PR, whose training would be much richer in examples of effective action, including mental action, might have a model capable of performing operations like this, and more elaborate iterations, spontaneously.

PR could develop such capabilities in more than one way. It could see other agents thinking aloud about their own recall problem, and apply the predictive model developed for their behavior to predict its own actions, by analogy. Once it has a predictive model of its own recall behavior in some contexts, it would extend that to other contexts, also by analogy.

Pulling these arguments together, we can suggest how a predictive model can produce behaviors, like answering questions, that we think of as "recall". This is true even though a predictive model lacks the simple store and retrieve structure we generally think of. A predictive model can also account for the complex search processes that are the most striking aspect of the Williams yearbook study. Now we'll move on to consider other evidence about the workings of human memory, and how a predictive model might account for it.

## Notes

(1) As we've discussed, LLMs are trained on large bodies of text. The training happens at a certain time, and after training is completed, the predictive model is never updated. That means that the LLM doesn't know anything about events that occurred after training.

As of February 2024, the training materials for chatGPT included only material available as of April 2023. So the system can't answer questions about results of the US Open tennis tournament that year, which happened in the summer... but wait! Let's try it:

```
USER
The women's singles title holder for the US Open is
ChatGPT
Based on a quick search, here's what I discovered.
The women's singles title holder for the 2023 US Open is Coco
Gauff
```

What has happened there is that chatGPT has looked up this information using the Bing search engine, rather than relying on its predictive model. Later versions of chatGPT include added capabilities like that, that can make it tricky to identify the capabilities it gets from its prediction training.

(2) ChatGPT's use of an external resource, a search engine in this case, of course mirrors human use of such resources. Arguably, the use of external resources, such as paper and pencil, as well as books and other references, are vital to modern human capabilities. There are many kinds of thinking that we can perform much better if we can make and refer to notes or calculations.

This brings us to the edge of an interesting topic in psychology and the philosophy of mind, the *extended mind* (see for example Rupert 2009). This work, that we won't be able to explore, raises the question whether the mind exists only in the brain, or should instead be considered to include things outside the brain that play key roles in thinking.

(3) Wang et al. (2022) describe a modified LLM that improves its performance by carrying out multiple predictions, using different reasoning paths, and choosing the response on which most paths agree. This is reminiscent of the consistency check carried out by Williams' participants.

# Interference with Real World Knowledge 10

***Focus:*** Lewis and Anderson (1976) Interference with real world knowledge. We discuss how knowledge is thought to be represented in human memory, in one influential account. PR works quite differently. Can it account for the results of an experimental study of human memory?

As we've seen, production systems work by modifying a memory. In the study discussed in Chap. 6, memory was used mainly to store information about the progress of a task: what the current goal is, and what step in a complex task has been reached. But in more general models of cognition, memory is used to store all kinds of information, including facts about the world, or memories of recent events a person has observed. Such memory is called *declarative*, as distinguished from *procedural*, memory. In a production system, procedural memory, knowledge of how to do things, resides in the productions themselves, not in the memory that production rules act on. We'll discuss the relationship between declarative and procedural memory further in a later chapter.

In this chapter, we'll discuss some influential ideas about how information is stored in declarative memory. We'll focus on an experiment in which people studied made-up information about historical figures and showed what learning the made-up information affected how quickly they could access real information about those people. As usual, our interest will be in whether or not a predictive model can account for these results. Along the way, we'll consider a new aspect of predictive models: how quickly do they perform different tasks?

ACT-R, a leading, very well-developed model of human cognition, is a production system. Information in its memory is represented as a network of elements called *chunks*. Figure 10.1 (from Anderson and Schunn 2013, p 3) shows a piece of network, representing part of what someone may know about numbers and arithmetic.



C. Lewis, *Artificial Psychology*, Synthesis Lectures on Human-Centered Informatics,
https://doi.org/10.1007/978-3-031-76646-6_10

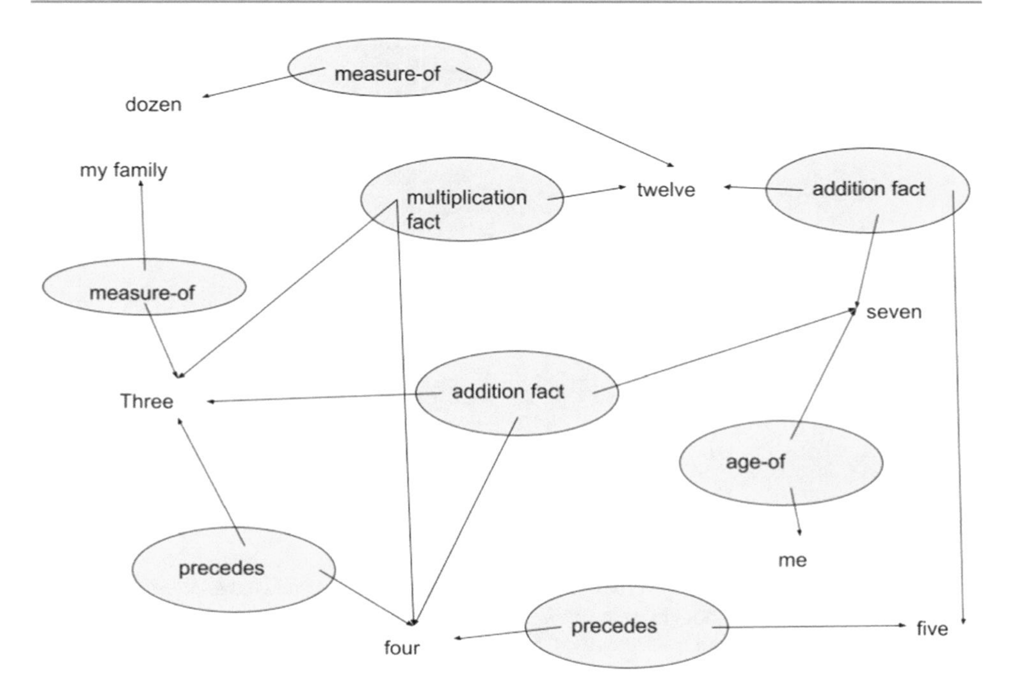


**Fig. 10.1** A network in memory

Don't worry about understanding all the details in this picture. Here are the key points.

First, there are different kinds of chunks shown. Some, like "My Family", "Three", and "Me", correspond to things the person knows about. Other chunks, shown in ovals, represent *relationships* among things that the person knows. For example, the "measure-of" chunk, shown at the top of the figure, represents the fact that the person knows that the number "Twelve" is a "measure-of" "Dozen". The large "addition-fact", near the middle of the figure, represents a relationship among Three, Four, and Seven: "Three" added to "Four" is "Seven". (The figure is simplified by not labeling the arrows that connect the chunks. In a full representation the arrows connecting the "addition-fact" chunk to "Three" and "Four" would be different from that connecting to "Seven", so that one can tell that "Three" and "Four" are the numbers to be added, while "Seven" is the answer.)

Second, by combining chunks of these kinds, we can represent things someone might know. For example, the person whose knowledge is represented in the picture has a family ("my-family") with Three members, since "measure-of" "my-family" is Three.

Third, productions can operate on a memory like this one, by testing for chunks that are or aren't there, and by adding new chunks if appropriate conditions are satisfied.

The conditions of productions specify chunks that may or may not match chunks in memory. The matching process is complex, in that conditions can specify multiple chunks, that have to be related in particular ways, and matches can be partial rather than complete (as mentioned in Chap. 6).

When the condition of a production rule is satisfied, its action may be applied. The action specifies new chunks that are added to memory.

Not all chunks are equally readily matched, depending on how often they have been retrieved, in the past, how recently, and how they are connected to other chunks. That is, chunks have a level of *activation*, and more active chunks are more quickly matched than less active ones. Also, activation spreads along the connections between chunks, so that (for example), if the node "Four" in the picture is highly active, say because it has just been used in a problem solving step, the addition and multiplication facts it is connected to, and its relations to "Three" and "Five" will become more active.

These ideas about memory come from a long tradition in psychology called *associationism*, that holds that ideas in memory are interconnected by associations that can be strong or weak, and that thought moves from one idea to another based on strength or association (we'll discuss these ideas further in a later chapter.) ACT-R has refined these ideas, adding the existence of different kinds of associations (illustrated by the difference between the connection from "addition-fact" to "Three" and "Four", and the connection to "Seven", in the piece of network we've looked at). Of course ACT-R has also added its procedural memory, the production rules, that prescribe the way the contents of memory are modified.

This idea of a network of chunks may seem like a lot of complexity. Couldn't memory just be a list of facts? Unfortunately, to be useful, facts have to be interrelated, and just listing them doesn't help with that. For example, I may know that my neighbor is Fred, and Fred drives a Tesla. To know what car my neighbor drives, I need to connect those facts. A network that has a chunk for Fred, and chunks relating me to Fred, as my neighbor, and Fred to his car, makes this easy. I can just follow the connections from me to Fred to the car.

The details of the matching process in ACT-R have been shaped by the results of a great many studies. Many of these have studied something called the *fan effect*. The general idea is that if activation spreads from chunk to chunk, as envisioned in the basic idea of association, it may get thinner, and slower, if it has to spread along more connections, connections that *fan out* from one chunk to others. If there's only one connection moving on from some chunk to another, more activation is transmitted, more quickly, than if there are ten onward connections.

Yours truly performed one such study, advised by John Anderson, in my grad student days. Previous studies of the fan effect had asked participants to memorize collections of made-up facts, devised so that the amount of fan in the connections among the chunks that could represent the facts was varied. Results were in line with the predictions of the spreading activation theory. But if human memory really has this network form, shouldn't fan effects also be seen for real facts, not just made-up ones?

In Lewis and Anderson (1976), participants were asked to learn "fantasy facts" about real people. That is, they were told that they would be given made-up "facts" about familiar public figures, and that they should write a brief continuation of each, that would relate in some way to their understanding of it. Example continuations in the instructions were adding details, and reconciling the "fact" with what they knew about the person. They were not told that they would be tested on their memory for the "facts", or the continuations, but that the continuations would be part of a study of how people work with fantasy worlds.

The fantasy facts were constructed by pairing a predicate from one real person with another real person, to whom it did not apply. For example, the predicate "IS A WOMAN", from the real fact "GOLDA MEIR IS A WOMAN" might be paired with MARK TWAIN to give the fantasy fact "MARK TWAIN IS A WOMAN". In one of the experiments in the study, participants saw no fantasy facts about some public figures, one fantasy fact for others, and four for other figures.

After a period of working with the fantasy facts, participants then were tested on how quickly they could recognize a sentence as true or false, indicating their response by pressing one of two keys. In one form of this test, they were told to respond TRUE only to actual facts, and no fantasy facts were shown. That is, each item was either an actual fact, or a false item. The false items were constructed using public figures and predicates like those used to create the fantasy facts, but not used in the fantasy facts, as well as those that had occurred in the fantasy facts, but paired differently. The false items were constructed in such a way that any given name, or any given predicate, occurred equally often in true and false items.

The key data from the experiment was the speed with which the participants could recognize an item as TRUE.

One can see the fan effect in the data (Table 10.1). If someone had studied no fantasy facts about (say) GEORGE WASHINGTON they could recognize GEORGE WASHINGTON CROSSED THE DELAWARE more quickly than if they had studied one fantasy fact about him (say, GEORGE WASHINGTON WAS A WOMAN) or four fantasy facts (say, GEORGE WASHINGTON WAS A WOMAN, GEORGE WASHINGTON IS CUBAN, GEORGE WASHINGTON WAS ASSASSINATED, and GEORGE WASHINGTON WAS FRENCH).

In ACT-R's model of declarative memory, this is expected. That's because in trying to match GEORGE WASHINGTON CROSSED THE DELAWARE activation from GEORGE WASHINGTON has to meet activation from the predicate CROSSED THE DELAWARE, to check that that person, and that predicate go together. When more fantasy facts have been added to GEORGE WASHINGTON, the spread of activation is slowed.

Results like these have provided evidence for memory being a highly structured network. Is there any possibility for accounting for such effects in a predictive model?

Right away we encounter a serious challenge. As of now, we have no way at all of predicting how rapidly a predictive model could operate! Existing LLMs provide no guidance. They work by carrying out complex matrix and vector arithmetic, and other mathematical

**Table 10.1** Fan effect on real-world knowledge

| Number of fantasy facts | 0 | 1 | 4 |
|---|---|---|---|
| Response time (msec) | 1360 | 1400 | 1440 |

operations, that aren't affected in any way by the specific model that's being evaluated. For example, if the same network architecture is trained on two different corpora, the resulting models will produce predictions in exactly the same amount of time, no matter how different or similar the two corpora are. Different corpora used in training will result in different values for the billions of weights in the model, but these are just different numbers. The time required to operate on these numbers doesn't depend at all on what the numbers are. Thus, it seems that a model that has learned more fantasy facts would operate at exactly the same speed as one that has learned fewer.

Nevertheless, there may be a way that a predictive model, even one closely based on existing LLMs, could produce responses in different amounts of time. If we think back to the Williams yearbook study, we saw that recalling a name is not a one-step process. Rather, it can consist of varying amounts of searching, in different contexts, and using different strategies. It also may have different amounts of checking.

Could the amount of processing involved in verifying an assertion about GEORGE WASHINGTON be influenced by the number of fantasy facts that have been learned? It seems plausible that it could be. A simple process for which this would be true is this:

```
to test if GEORGE WASHINGTON CROSSED THE DELAWARE:
think of something GEORGE WASHINGTON did
if the thing thought of is CROSSED THE DELAWARE say YES
otherwise try to think of something else GEORGE WASHINGTON
did, and repeat
if you can't think of something else GEORGE WASHINGTON did,
say NO
```

The more somethings that GEORGE WASHINGTON did, that the system has learned, the longer this process will take to respond.

Undoubtedly, this process can't be correct. For one thing, it doesn't allow for processing that includes asking who CROSSED THE DELAWARE, as well as asking what GEORGE WASHINGTON did, as a way of determining whether or not GEORGE WASHINGTON CROSSED THE DELAWARE. It's hard to see why that wouldn't be done. But it does suggest how a system that performs its basic operations at a fixed rate could generate responses at a variable rate, depending on what it has learned.

It seems that a predictive model could carry out this process. Here is an example, from GPT-4:

```
Fred likes beets. Fred likes carrots. Fred likes cabbage.
Give me an example of something Fred likes:
ChatGPT
Fred likes beets, carrots, and cabbage. An example of
something Fred likes is carrots.
```

```
User
Is that example"broccoli"?
ChatGPT
No, the example I provided was "carrots." However, based on
the information given, Fred likes beets, carrots, and
cabbage. There was no mention of Fred liking broccoli.
User
Give me a different example of something Fred likes.
ChatGPT
Fred likes cabbage.
User
Is that example "broccoli"?
ChatGPT
No, the example I provided was "cabbage." Broccoli was not
mentioned as something Fred likes.
User
Give me a different example of something Fred likes.
ChatGPT
Fred likes beets.
User
Is that example "broccoli"?
ChatGPT
No, the example I provided was "beets." Broccoli was not
mentioned as something Fred likes.
User
Give me a different example of something Fred likes.
ChatGPT
Based on the information provided earlier, Fred likes
beets, carrots, and cabbage. Since I've already mentioned
beets, carrots, and cabbage as examples of what Fred likes,
those are the only examples available. There's no
additional information suggesting Fred likes anything else,
such as broccoli.
```

We can see that the system can do the following things:

- It can retrieve "facts" about Fred.
- It can determine if a retrieved "fact" matches a probe, "broccoli" in this case.

- Having retrieved some facts, it can retrieve a fact it hasn't already retrieved, if there is one.
- It can determine when there are no more facts to retrieve.

These are all the operations needed to carry out the search process in question.

You may wonder why the example uses "Fred" and not George Washington. The answer is that the GPT models are too good at keeping real facts and fantasy facts separate. When asked to come up with facts about George Washington, they only give real facts.

That's more reason to doubt that the search process above is what participants in the Lewis and Anderson study were doing. Nevertheless, the process serves to demonstrate that the fan effect, in which more fantasy sentences lead to slower responses, can occur in a system that does not have a memory structured as a network. It also shows that variation in response time can occur even in a system whose fundamental speed of operation is not affected by what it has learned. That's because different bodies of learned materials can call for different numbers of processing steps.

Taken together, these arguments suggest that predictive models, that do not have structured memory, or indeed any memory other than the predictive model itself, might be able to account for the fan effect. This is true despite the fact that the fan effect is often taken as a signature phenomenon for network structure.

There is much more to be said about speed of responding as data in psychological studies, however. In the next chapter, we'll consider the phenomenon of the speed–accuracy tradeoff, in which people are able to manage their performance on many tasks so as to produce accurate results, slowly, or less accurate results, more quickly. That is, speed of responding is shaped not only by the materials to be processed but also by a wide range of factors, including whether participants are told they should work carefully, or work quickly. How can this flexibility be explained?

# 11 Speed–Accuracy Tradeoffs

***Focus:*** Howell and Kreidler (1963) Information processing under contradictory instructional sets. How do instructions to work quickly or carefully influence what people do? Can PR account for this?

We've seen in the last chapter that how fast someone can respond to a task can depend on the materials they have to work with, fantasy and real facts, in the focal example in that chapter. It's long been noticed that speed of responding can be influenced by many other things, including participants' efforts to respond accurately, or to respond quickly. The phrase *speed–accuracy tradeoff* refers to the fact that, in general, responding more accurately usually takes more time, while responding more quickly produces more errors. In this chapter, we'll explore how a predictive model might account for this.

Howell and Kreidler (1963) performed an early experiment that demonstrates this, using explicit instructions to influence how participants behaved. They used an extremely simple task. Participants saw a row of 10 lights, with a button immediately below each light. During the experiment, one of the lights would light, and the participant simply had to press the corresponding button.

The point of the experiment was in the instructions participants received. One group was told (Howell and Kreidler 1963, p 41),

> You are to respond as fast as possible; . . . although you would like to make correct responses, speed is the important thing and accuracy is definitely a secondary consideration.



C. Lewis, *Artificial Psychology*, Synthesis Lectures on Human-Centered Informatics,
https://doi.org/10.1007/978-3-031-76646-6_11

**Table 11.1** Speed–Accuracy tradeoff

| Group: | Speed | Accuracy |
|---|---|---|
| Responses/sec | 1.84 | 1.72 |
| % Correct | 87.5 | 97.5 |

Another group was told,

> You are to respond so as to make as few errors as possible; . . . speed is definitely a secondary consideration.

We'll call the first group the Speed group, and the second the Accuracy group.

The results are shown in Table 11.1.

We can see that the Speed group is faster than the Accuracy group, and the Accuracy group is more accurate than the Speed group.

These results likely aren't surprising. After all, people are just doing what they were asked to do. The important question is, what are the groups doing differently to produce these results? A companion question for us is, is there any way a predictive model could show this flexibility?

There's a vast literature on the first question, with many different models proposed, evaluated against data, and (in some cases) rejected, and in others, kept under consideration. To illustrate, one class of models assumes that evidence for or against different responses is accumulated over time, until some *threshold* is reached for one of the responses. The evidence is assumed to arrive randomly in time, so that only by waiting forever could one be completely certain of a response. To respond more quickly, but less accurately, one would set a low threshold for responding. That way the threshold will be reached soon, when not much evidence has accumulated, so that errors are pretty likely. To respond more accurately, one sets the threshold higher. That way more evidence has to come in, and the response will happen later, but be more likely correct.

In thinking about this in a predictive model, we face a situation quite parallel to that for the fan effect. Current LLMs, using the transformer architecture, do not have an accumulation process in them. As we saw in the earlier discussion, their basic operation time is fixed, and there is no threshold for responding that can be modified.

Another way to produce a speed–accuracy tradeoff can be envisioned, however, parallel to the kind of thing proposed for the fan effect. Different speeds of operation can be produced, if a decision process is used that can take varying numbers of prediction steps. To obtain a speed–accuracy tradeoff, we'd imagine that responding under the accuracy condition would require checking operations, for example, looking again at the lamps and buttons. These operations would not be used, or perhaps might be used less often, when given the speed instructions.

This approach has the attractive feature that it answers a question that is usually not asked when these matters are considered: how does it happen that particular instructions to a participant lead to the participant behaving differently? If we're imagining an accumulator model, for example, how is it that participants set their threshold in a particular way? No one thinks participants are aware that there is a threshold, or an accumulation process; if they were, it would be much clearer what's actually happening. So how is it that they set it appropriately, given particular instructions?

In a predictive model, we can readily answer that question. To interpret instructions, and act on them, just is to predict appropriate actions, given the instructions. These actions are then the ones that are carried out. Participants' predictive models reflect the regularities that connect various instructions with various response processes, for example, checking or not checking.

This line of thought also accounts for other facts. Instructions are one way to influence how participants respond, but there are others. For example, participants may be paid something for correct responses, and penalized for errors. By making the penalty for errors small or large, relative to the reward for correct responses, one can push participants towards speedy or careful processing. Or one can impose *response deadlines*, in such a way that only responses made within a certain time are rewarded, as a way to promote speedy over careful responding. In a predictive model, all of these influences would be explained by predictive regularities that connect elements of the instructions, as well as things seen during the experiment, with various processing steps.

This flexibility is possible because the predictive model integrates everything an agent "knows", rather than separating it into multiple collections of knowledge, whose interaction can be problematic. In particular, things the agent "knows" about the meanings of words, including what "speed" is, what an "error" is, and so on, are embedded in the same model that produces actions. Actions that are produced are those that are predicted, given the predictive force of the words, influenced by the language patterns the words occur in. There are no hidden mechanisms, like thresholds, that have to be controlled.

While this picture may have its attractions, there's good reason to doubt that it can be correct. At least, it seems likely that there are considerable complications in covering the whole landscape of speed–accuracy tradeoff along these lines. A review by Heitz (2014) cites diverse settings in which speed–accuracy tradeoff has been observed, including a study by Chittka et al. (2003) on bumblebees. In their study, bees were foraging for sugar solution in a simulated flower meadow, in which color patches represented "flowers". A small color difference distinguished rewarding flowers, ones that had some sugar solution available, from unrewarding flowers, that offered only water.

The data showed a speed-accuracy tradeoff among the bees: bees that foraged more slowly visited fewer unrewarding flowers. Also interesting for us is what happened when the water in the unrewarding flowers was replaced by quinine solution, something bees don't like. Individual bees shifted their foraging so as to take longer, and make fewer mistakes. That is, a larger penalty for mistakes led to more careful, slower foraging.

We can't suppose that the way humans respond to verbal instructions, for example by checking or not checking, has anything to do with how the bees are responding. It could perhaps be that predictive connections in the bees' experience does play a role, but how would we imagine their visual discrimination is being affected?

Before leaving the topic of speed–accuracy tradeoffs, let's note that there is a direction of inquiry that we haven't explored. We've discussed just one kind of predictive model, based closely on existing LLMs. As we've seen, these models have limitations in explaining variations in response time, because their basic speed of operation is fixed. Towards the end of our exploration, in Chap. 26, we'll consider predictive models that do not work that way, and may offer other accounts of response time data.

For now, we'll continue our exploration of memory by examining theories of how knowledge is represented in human memory. As we'll see, a predictive model works quite differently from these.

# 12 Is Knowledge Represented by Propositions?

***Focus:*** Influential psychological theories propose that information in memory is represented by structures called propositions. What is the evidence for this? PR doesn't represent things this way. Can it be squared with the evidence?

In many psychological theories, facts are represented as *propositions*, which relate entities that refer to objects, or to abstract entities, or other propositions. A proposition is taken to represent the meaning of a sentence, like "The sum of 3 and 4 is 7", which is one of the examples discussed in Chap. 10. In fact, a proposition represents a whole family of sentences, not just that one, but also "7 is the sum of 3 and 4", and many others.

Propositions play an important part in many theories. Is it possible for a predictive model do without them? To explore this matter, we'll need to spend some time on what propositions are, and how they are used in psychological theories. There are many different theories that use them, and I'll be illustrating them here, by showing some different ways propositions are represented. But please don't get bogged down in the details of the examples. They are just intended to give you a sense of how propositions are represented. Our concern is only that all of these theories use propositions in a similar way, without being concerned with the differences among them.

I've said that propositions can represent the meaning of sentences. But just what sentences are represented by a particular proposition is difficult to pin down—is the sentence "The sum of 4 and 3 is 7" represented by the proposition we're imagining here, or a different one? Does our representation of sentences about sums assume that we know that summing is commutative? We'll pass over this matter and go forward on the assumption that we have some way to translate sentences into propositions, in such a way that sentences that mean the same thing will be translated to the same proposition.



C. Lewis, *Artificial Psychology*, Synthesis Lectures on Human-Centered Informatics,
https://doi.org/10.1007/978-3-031-76646-6_12

The fact that sentences with the same meaning are represented by the same proposition is one reason to use propositions to begin with. Doing this provides an easy explanation of how it is that, when we have told someone a fact, in any of a number of synonymous ways, we expect them to answer questions in a way that doesn't depend on the particular expression we chose for the fact. In a theory in which what is stored in memory is propositions, this happens because they have represented what we told them in the same way, regardless of how we told it to them.

Propositions are often decomposed into relationships between other entities. A common form relates a subject and a predicate, where a predicate may relate a verb and an object. Thus the sentence "Pat kicked the ball" might be represented by a little patch of network like this (Fig. 12.1): We can see here that the sentence is actually represented by two propositions, not just one. One proposition says, in effect, that Pat kicked something, and the other proposition says that the something is a ball. Each proposition is connected to its subject and predicate by S and P arrows. The W arrow connects a node to a word that can be used to describe it. Kicking is represented by a node that has an R arrow connecting it to a relation, and an A arrow connecting it to an argument, the node that represents what Pat kicks. In the example, that node is the subject of the proposition that says it's a ball.

The details here don't matter. We just need to see that there are propositions being used to represent the meaning of the sentence.

There are many different representational schemes of this kind. This example follows Anderson (1976, p. 151). The same sentence, represented as in Van Dijk (1983, p. 114), is shown in Fig. 12.2.

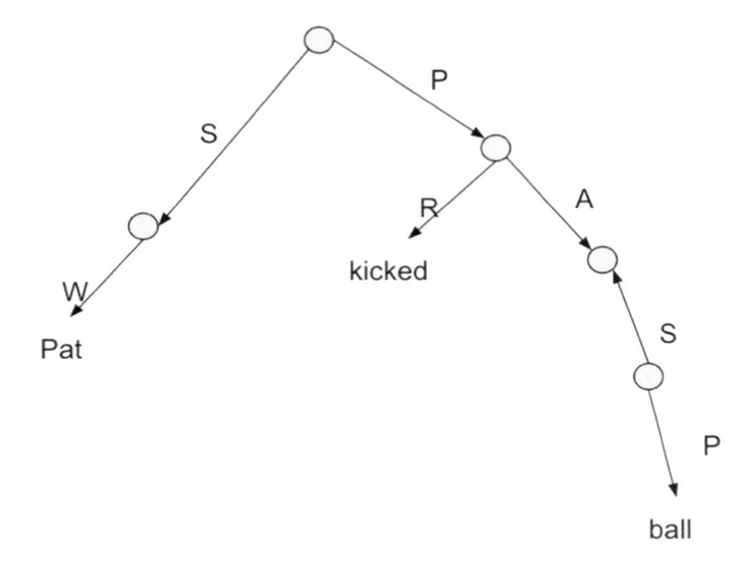


**Fig. 12.1** A network representation of a sentence

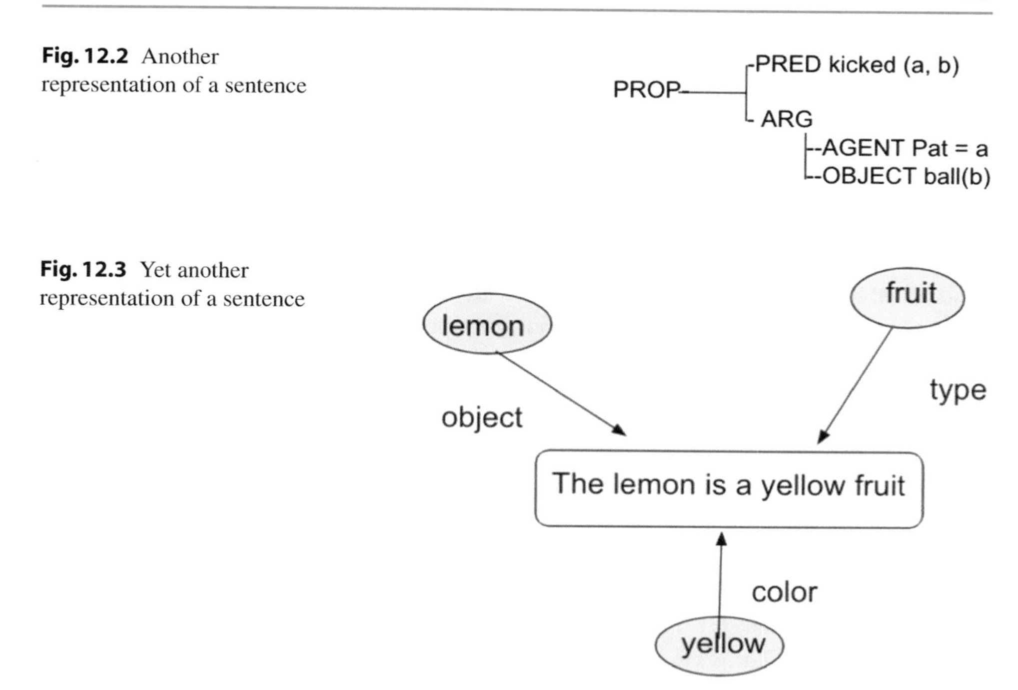


**Fig. 12.2** Another representation of a sentence

**Fig. 12.3** Yet another representation of a sentence

This looks quite different from the network we looked at first, but we can see that the same information is being represented. It might appear that there's only one proposition, called out by the PROP at the left, but there's a second proposition expressed as ball(b) in the description of the OBJECT. So there are two propositions here, as in the first network. Connections between entities in the network look different, since arrows aren't used, but the same information is there. The subject, "a", is given the name "a" in the expression for the AGENT. The object of the kicking, "b", is described in the OBJECT as being a ball, that is, something for which being a ball is true.

The examples we've looked at are drawn from quite old sources, but representational schemes like these are still current. For example, Fig. 12.3 is adapted from Stocco et al. (2024, p. 134).

Note here that the "The lemon is a yellow fruit" is just a label for the node in the middle, and isn't part of the representation. That node represents the idea that is specified by its connections to the other nodes shown: it's an idea about an object, lemon, that attributes to it a type and a color.

As I've said, the differences among these representations, and their details, need not concern us, because our interest is, why use this kind of representation at all? We've already mentioned one reason, that representing sentences by propositions allows us to represent a given body of information in the same way, even when it is presented differently, for example, in different languages.

Another reason is that representing the meaning of sentences in this way allows us to form connections among parts of different sentences, as we saw in the chapter on Interference. If we are given a text with many, related, sentences in it, and we represent them all using either style of network, we'll end up with many pieces of network stitched together. For example, we might have another sentence that tells us that the Pat who kicked the ball is a lawyer, or that the ball was large and heavy. Representing these sentences gives us additional propositions that would be connected with those we have constructed to represent the first facts about Pat, and what Pat kicked.

Theories of this kind go on to specify how networks like these are built, when someone reads a text, for example. They also go on to describe how questions can be answered, as we saw in the Interference chapter, or how logical inferences could be drawn from a collection of propositions.

To the extent that one can describe processes like these, one can feel that propositional representations can help explain things people do, in processing information. They suggest ways these processes might work. But how confident can we be that that is actually how people do them? Do we really have propositions in our heads?

We've already discussed one line of argument about this, in the chapter on interference. It's possible to account for many things about how fast people can access information in their memories, using this kind of representation.

Another line of argument looks for evidence that the patterns of connections in these networks reflect real connections, that is, that things that are assumed to be tightly connected in these networks are actually tightly connected in people's memory. One way people have tried to examine this question is by asking people to read sentences, and then remember whatever words from the sentences they could. The idea is that words that were closely linked in these networks would be recalled together. Another technique used *cued recall*: one could present one word from a sentence, as a cue, and ask participants to recall other words. Here the idea was that closely linked words would be good cues for one another. Unfortunately, the results have turned out mixed, and difficult to interpret (see Anderson 1976 for an early discussion.)

A related question is whether propositions act as indivisible units, or whether pieces of them can be recalled. A complication in exploring this is that it appears that what people remember, after reading a text, isn't just the meaning of the text, which would be captured by one or more propositions, but can include the specific words used. That is, people may remember the sentence that expressed a proposition, as well as the proposition.

Anderson (1983) concluded that there wasn't convincing experimental evidence that propositions were indivisible, all-or-none, units, but that there were computational reasons for assuming that they were indivisible:

> The empirical evidence provides weak evidence at best on all-or-none memory for propositions ⋯ . In contrast to the murkiness of the empirical picture, the evidence for an all-or-none system is quite clear from considerations of functional value within a production system. Our experience has been that it is not adaptive to store or retrieve partial

> propositions, because partial information cannot easily be used in further processing, and its retrieval only clutters up working memory, or worse, misleads the information processing [pp. 72–73].

Recall that Anderson uses production rules to represent procedural knowledge, and the network to represent declarative knowledge. For declarative information to be retrieved is for it to be matched by the condition of some production, whose action can then do further work, like saying something, or drawing an inference. So Anderson is arguing that it's difficult to get a collection of productions to work properly, if their conditions are allowed to match parts of propositions.

In later work on the same family of models, however, partial matches are permitted (see e.g. Stocco et al. 2024). Other changes to the theory have apparently moderated the difficulties with this.

It seems, then, that firm evidence for our having propositions in our heads is not available. But is it possible to do the work propositions are thought to do, in answering questions or drawing inferences, without them? Here is Anderson's summary of his motives for using them:

> Propositions have three important features– they are abstract, have truth values, and have rules of formation. Their abstractness means that our declarative knowledge is not tied to the modality, language, situation, etc., of acquisition. So we can report what we learned in French, or write about we have seen. The feature of having a truth value allows all our declarative knowledge to enter directly into inferential processes· · · . The rules of formation enable us to relate network structures to their meanings · · · Anderson (1976, p. 120).

We've already discussed the last point, about rules, in the Einstellung chapter. In agreement with Harold Garfinkel, we suggested that rules are generally not a good way to capture meaning, outside artificial contexts. On Anderson's first two requirements, we can see immediately that predictive models can meet them, without any use of propositions at all. Indeed, current LLMs, with none of the extensions proposed for PR, can show this.

For abstraction, consider this conversation:

```
User
Tenga en cuenta, por favor: Fred y Ethel son dos de mis
personajes de televisión favoritos.
ChatGPT
Fred y Ethel son personajes muy queridos de la icónica
serie de televisión ''I Love Lucy'', que se emitió desde
1951 hasta 1957. ... Su amistad y las travesuras en las que
se embarcan con Lucy son una parte fundamental del
encanto y el éxito duradero de ''I Love Lucy''.
```

```
User
Can you tell me about one of my favorite TV characters?
ChatGPT
Absolutely, I'd be glad to! Please tell me the name of the
character and the TV show they're from, and I'll share some
information about them.
User
I already told you.
ChatGPT
My apologies for the oversight. Since you mentioned Fred
and Ethel from ''I Love Lucy'' as your favorite TV
characters, let's dive into one of them more deeply.
How about we focus on Ethel Mertz? ... Ethel's character is
essential to the dynamic of ''I Love Lucy,'' serving as a
foil to Lucy's more flamboyant and impulsive nature. ...
Is there anything specific you'd like to know about Ethel
or any particular episode or moment involving her that you
love?
```

Here I have provided some information to GPT, in Spanish. I then ask a question in English that draws on that information. GPT at first doesn't recognize that I am asking about what I've already said. But when I correct it, it answers in English, giving a response that shows it had no trouble making use in English of the information I provided in Spanish.

Examples of inference are also easy to find:

```
User
If Denver were east of Boston, what direction would I have
to travel to get to Boston from Denver?
ChatGPT
If Denver were hypothetically located east of Boston, you
would have to travel west to get to Boston from Denver. In
reality, Denver is west of Boston, so typically you would
travel east to go from Denver to Boston.
```

Note that because this is an inference from a false premise, the answer can't be obtained just by retrieving facts about the two cities. GPT has to make an inference.

Does that mean the GPT has propositions, since it can do those things?

The issue of *virtuality* arises here, as it has before, and we'll return to it later. When something acts as if it has something, there's always the possibility that it has that thing, virtually. But we can note that nothing in the construction or training of GPT made explicit use of the idea of proposition in any way. There is no data structure in the program that was designed to represent propositions, or that was shaped by any ideas about propositions that the designers might have had.

This contrasts sharply with the models we've been discussing. In those, there are data structures that are expressly designed for representing propositions, and that happened because the theorists were responding to ideas they have about propositions and their roles.

We see that systems that were developed with no provision for propositions in their internal structures, and with no thought of them in the design process, succeed in meeting the functional requirements that have motivated psychologists to propose that people have propositions in their heads. That success suggests that we should reevaluate the argument. Since it appears that we *need not* have propositions in our heads, perhaps we *do not* have them in our heads.

The possibilities in this rethink go beyond just doing without propositions. The idea that what we have in our heads is not a collection of propositions, but rather a predictive model of our experiences makes it easier to understand how what's in our heads relates to our experiences.

When we know that "Pat kicked the ball" we usually know indefinitely more than just that. We may know who Pat is, where this happened, what kind of ball it was, where it went, and much more. A scheme that tries to represent this information, explicitly, has to have a place to specify all these kinds of things. While it is possible to elaborate the representational scheme to make room for all this, how are the connections that are needed actually established? That is, how does one process a visual experience, or a text, so as to find all these relationships, and build the network structure that captures them?

No doubt some kind of story can be told. Computer programs, on which these theories are modeled, and by which they are simulated, are extremely flexible. But to build this kind of theory a lot of programming has to be done, programming that is special to all of the specific ways things in propositions might be specified. For example, the Kintsch and van Dijk theory has locations, and modes of action, and times, and so on. Each of these has to be managed by particular machinery, to extract relevant information from text (for which a good deal of theory has been spelled out) but also to extract relevant information from the world, for which not much has been attempted. Each kind of information calls for a different story to be written.

In a prediction model, by contrast, all and only the information that has predictive value is represented. The principles of representation are the same, no matter what kind of information is concerned.

There's a parallel in the processing of language by LLMs. Linguistic theorists have often imagined that the structures we see in language have to be reflected in structured representations in the head. Complex schemes have been devised to capture structures of different kinds, with difficult analytical work required to dope out apparently relevant structures, and ways to represent them, compatible with other ideas about structure.

But, as we'll see, LLMs achieve command of many aspects of language, with no one having to do any of that analytical work at all! The demands of prediction, coupled with a sufficiently flexible, and extremely simple, representational scheme– sequences of vectors– suffice to produce command that goes far beyond any of our prior attempts.

Further, and in line with what's needed in our account of memory, the command of language is combined with command of many aspects of "real life": purposes, professions, materials, structures, and so on. Here again, no analytical work whatsoever was needed to contrive this. We'll consider these matters further in a later chapter. For now, we'll continue our consideration of theories of human memory, and alternatives to them suggested by predictive models.

## Note

For philosophers, propositions are abstract entities that can be discussed without reference to any particular way of representing them. In a parallel situation, we might distinguish the number 2 from any particular way of writing it, as a digit, as just done, or by the Roman numeral II, or by an arrangement of charges in a computer memory, for example. Our discussion has addressed propositions as represented in (many) psychological theories, in which one can point to a piece of semantic network, and say, "There, there's where that proposition is". It follows that in suggesting that propositions can be dispensed with, we are saying something about psychological theories or models, and not about propositions as abstract entities.

# Associationism 13

***Focus:*** How predictive modeling relates to an old and enduring idea in psychology: "association" between ideas or thoughts. Does predictive modeling evade criticisms to which associationism has been subjected?

An idea in psychology traceable as far back as Aristotle is that the flow of thought is driven by *associations* between ideas. Associations arise from *contiguity*: when one thought follows another, they become associated, with a tendency for the re-occurrence of the first idea tending to produce a re-occurrence of the second idea.

This sounds much like the prediction framework we've been discussing, but if it is, we could be in trouble. Long ago, associationism was subjected to a critique that many found persuasive.

As reviewed by Walter Kintsch (1974), and excerpted in Mandler and Mandler (1964), German Psychologist Otto Selz argued in 1927 that there need to be different kinds of associations between ideas, not just a single contiguity-based association.

For example, suppose one is asking a test participant to name a category to which a probe word, presented a moment later, belongs. The instruction to name a category would, it was thought, activate a collection of category names, including "occupation", and "tradesman". If one then presents "farmer" as the probe, simple association would make "occupation" a likely response, because (1) it's a category name, and hence associated with the instructions, and (2) it's associated with "farmer", because "farmer" is an occupation. So far so good.

But "tradesman" is also a likely response. (1) It's a category name, and hence associated with the instructions, and (2) it's associated with "farmer" because it occurred along with "farmer" on a common tax form in the Germany of the day. The problem is that "tradesman" isn't actually an appropriate response, since it isn't a category that "farmer" belongs to.


C. Lewis, *Artificial Psychology*, Synthesis Lectures on Human-Centered Informatics,
https://doi.org/10.1007/978-3-031-76646-6_13

To avoid this, Selz argued that the simple idea of association needed to be replaced by specific relationships. Thus a "superordinate" link would allow "occupation" to be picked out, given "farmer". As we saw in the previous chapter, this idea has been picked up in subsequent work on network representations of knowledge, that include different kinds of connections, or indications of the role a given entity plays when connected to another.

As we've seen, PR doesn't have this kind of connection between entities. Rather, Selz's problem is dealt with by *context*. In Selz's reconstruction of the "farmer" task, the effect of the instructions to choose a superordinate is just to activate candidate response words. The instructions don't directly influence the response to "farmer", leading to the difficulty he points out.

In a predictive model, the instructions continue to be part of the context for what the participant does, for as long as they are relevant. So the selection of a response for "farmer" isn't just that, but rather the selection of a response to "say a superordinate for ... farmer". Both "superordinate" and "farmer" are in the context, and they jointly influence the response.

This not only enables the predictive model to deal with Selz's problem but also makes it extremely flexible. In a semantic network representation, there are only so many kinds of connections, or possible roles for elements of memory structures. In an LLM, a context can be any sequence of words. For PR, with its expanded flow of experience, context can include any other experiences, including inner speech.

Another way predictive models differ from associative models and their descendants is in what the elements that are connected are. Traditionally, the elements were ideas, images, or (for William James), "things". In the more recent models, we've discussed, they are smaller meaningful entities, such as words. But in an LLM (and in PR) the elements *are not meaningful* at all.

In GPT, as we've discussed, the elements are tokens, many of which form parts of many words, and do not carry identifiable meanings. "Meaning" arises from the relationships among these meaningless entities that are found to have predictive value.

The next, and final, stage in our consideration of memory takes up a distinction that many theories have maintained, between memory for facts, and memories for skills.

# 14 Procedural and Declarative Knowledge

***Foci:*** Ryle, G. (2009) *The concept of mind*, and Cohen and Squire (1981) Retrograde amnesia and remote memory impairment. Is knowing a fact the same kind of thing as knowing how to do something? Can PR account for the possible differences?

We use the same verb in the sentences "Pat knows that the capital of Colorado is Denver", and "Pat knows how to ride a bicycle". But do these sentences describe the same kind of knowing? Philosopher Gilbert Ryle emphasized some differences between two kinds of knowledge, knowing *that* and knowing *how*, in an influential 1949 book (Ryle 2009). He noted that "knowing how" is graded, since one can do something more or less well, while "knowing that" seems to be all or nothing; that one can quickly learn that something is true, just by being told, while knowing how to do something usually develops gradually; and we expect that when someone "knows that" something, they can express it in words, while this is very often not the case for something they know how to do.

Theorists in psychology have responded in different ways to Ryle's distinction, commonly framed as a distinction between *declarative* knowledge (knowledge that one can declare, knowing *that*), and *procedural* knowledge (command of the procedure for doing something, knowing *how*.) We've seen that production systems, like Anderson's ACT-R, often make a sharp distinction between declarative knowledge, represented in the memory on which production rules act, and procedural knowledge, represented in the production rules themselves.

With this separation in place, it's not hard to account for Ryle's distinctions. Knowing a fact just requires having an appropriate structure in memory, which one has or does not have, while knowing how to do something requires having a collection of productions that coordinate in an effective way. Adding a new structure to memory can be done more or



C. Lewis, *Artificial Psychology*, Synthesis Lectures on Human-Centered Informatics,
https://doi.org/10.1007/978-3-031-76646-6_14

less directly, for example by interpreting a sentence. But learning a skill involves building production rules, for which there is no public form that can be just said by a teacher. Structures in memory can be examined and reported, by productions. But productions can't operate on productions themselves, so productions can't be reported.

Other theories, while not denying that there are differences between declarative and procedural knowledge, do not assign them to different storage systems. On the one hand, the LNR theory proposed by Norman and colleagues (Norman et al. 1975) envisions a networked memory, much like that in ACT-R, that stores *both* declarative knowledge, *and* structures that describe actions, and thus play the role of productions in ACT-R.

On the other hand, it's also possible to store declarative knowledge in the form of production rules, as described by Donald Waterman (1975). For example, instead of storing arithmetic facts like $7 + 2 = 9$, a production system can represent a way of calculating sums. (Note that a production system needs at least a little bit of memory, to keep track of the state of a process. But this need not be large enough to contain all of its declarative or procedural knowledge.) Waterman proposed that new knowledge is stored by creating new productions.

How do these unified schemes, ones that don't separate declarative and procedural knowledge, measure up against Ryle's distinctions? The first two distinctions, all or nothing vs. partial knowledge, and lack of direct instruction for procedures, can be handled the same way as for ACT-R. That is, the crucial distinctions needed aren't in where or how the information is stored, but in the nature of the information: is it one thing, or many? Is there an available public representation, that someone could use to teach it?

The reportability contrast takes some more work, as Anderson (1976) notes for LNR, in which facts and actions are described in the same network. How come some things in the network can be reported, while others can't be? Similarly, if all knowledge is represented as production rules, how does it happen that some of it can be reported, and some of it can't?

It seems likely that this challenge could be met. One approach would be to note that some nodes in memory have associated words, while others do not. For example, we have words for shoe and shoelace, so we can say things like "My shoes have shoelaces", but for the nodes that represent the actions involved in tying shoes, we don't have words, or at least, not words specific enough to provide an adequate description of the process.

What does all of this have to do with PR? Since PR has only its single, predictive model, one might say that PR is just another one of these single memory systems, that doesn't store declarative and procedural information separately. But there is more to be said. In fact, we can argue that not only does PR not have both declarative and procedural knowledge, stored together, but also that it actually doesn't store either one!

Of course, it *acts as if* it has both declarative and procedural knowledge. For declarative knowledge we can easily see such acting, in existing LLMs. We can ask GPT what the capital of Colorado is, and it will answer correctly. Unlike PR, GPT does not have an action system, other than one for saying things, so we can't assess its procedural knowledge very generally. But we can assess its procedural knowledge using prompts that require it to do things that are needed to say things. For example, to say the right thing in response to the following prompt, the system has to know how to reverse a sequence of items:

```
The reverse of the list orange apple lemon is the
following:> lemon apple orange
```

In what sense then, does GPT not have declarative and procedural knowledge?

We can start by contrasting it with ACT-R. ACT-R's declarative memory is a network of nodes that represent things and relationships. GPT has only its predictive model, which represents relationships among tokens. Crucially, it has no ability at all to inspect or report on its predictive model, showing that its predictive model just doesn't meet Ryle's definition for declarative knowledge.

Another perspective is that there is no part of the predictive model that we can point to, and say, "There, there's where the capital of Colorado is represented." We can get GPT to say what the capital of Colorado is, but it is not producing that answer by reading something stored in a declarative memory. Its predictive model produces answers, but not because it has stored them.

There's a slippery issue here of *virtuality*, as we encountered before. If a system behaves as if it has stored something, isn't it legitimate to say that it actually has stored it, that it has *virtual* storage? The answer is, yes and no. Yes, it can make sense to say that the thing has been stored. But no, if we assume that that means that our ordinary ideas about storage apply in this case!

For example, our familiar notion of storing something is that the thing stored can be located, intact, in the store, and that it is more or less distinguishable from other things in the store. Those things are true of the memory networks in ACT-R and LNR, but not for an LLM (and for many other ways of "storing" information). We'll return to questions of virtuality in Chap. 28.

How about procedural knowledge in GPT? We don't expect that knowledge to be reportable, and it isn't. But, as for declarative knowledge, we can't point to something in GPT's predictive model, and say, "There's where the knowledge of how to reverse a list is stored."

Note that it can look as if GPT "knows" procedures, as well as being able to execute them:

```
User
How would you reverse a list?
ChatGPT
You can reverse a list in several ways in Python. Here are
the most common methods:

1. Using the reverse() Method
```

```
The reverse() method directly modifies the original list
to reverse the order of its elements.

python
Copy code
my_list = [1, 2, 3, 4, 5]
my_list.reverse()
print(my_list)  # Output: [5, 4, 3, 2, 1]
2. Using the Slicing Technique
You can reverse a list by using the slicing technique.
This method creates a new list in the reverse order without
modifying the original list.
... bla bla bla
Each of these methods has its use cases depending on
whether you need to modify the original list or create
a new reversed list.
```

It does indeed "know" quite a bit about reversing lists, in that it can say quite a lot about the topic. But, as is given away directly in this answer, it doesn't know anything about how *it* would reverse a list. It's not a Python program, so that what we've seen it report isn't knowledge about itself!

This raises an issue that Ryle was concerned with, along with knowing that and knowing how. Some philosophers have argued that one can't know how to do a thing without knowing *that* something is true, and indeed without processing the *that* in some way. In particular, they say, any intelligent action must involve engagement with some fact or facts about appropriate action. Ryle argues to the contrary that "when we describe a performance as intelligent, this does not entail the double operation of considering and executing." Ryle (2009, p. 29).

Thus, we see that GPT reverses lists, but it does not follow that it "knows how to reverse lists", in the sense that it "knows" a procedure for doing it, or that its performance involves consulting and interpreting this knowledge.

In this way, GPT differs from LNR and Waterman, as well as from ACT-R. Those systems all, in different ways, represent "procedures". GPT represents them only virtually, as we saw is also true for facts.

Long after Ryle, interesting new data entered the discussion of the declarative-procedural distinction. In a 1980 paper, Cohen and Squire (1981) studied people with amnesia who were practicing the task of reading words shown in mirror writing. For example, the word "eggplant" would be shown as in Fig. 14.1.

**Fig. 14.1** "eggplant" in mirror writing

The participants all had major memory problems, as shown in a simple memory test. They were shown pairs of words, and then just the first word, and asked to remember the second word. The participants could only do this about 2 times out of ten after seeing a pair three times. People used as controls could get this 9 or 10 times out of ten.

If we think of the amnesic participants as having "trouble learning things", we might expect that they could not learn how to read mirror writing. That is, we might think that they would be just as slow in reading the mirror writing after a lot of practice as they were at first. In fact, though, the amnesic participants did improve with practice, showing that they could learn the skill of reading mirror writing. Also, their skill was retained for many weeks, as shown by the fact that their reading speed on a delayed test was much higher than at the beginning of the study, and about the same as before the delay.

So the situation is that the amnesic participants had great difficulty remembering words, which would require declarative memory (remembering facts about what words had been seen when), but did well learning a skill, mirror writing, that would be attributed to procedural memory. Cohen and Squire, and many researchers since, interpret these results as supporting the idea that declarative and procedural memory must be different:

> This distinction between procedural or rule-based information and declarative or data-based information, which is reminiscent of the classical distinction between"knowing how" and "knowing that," has been the subject of considerable discussion in the literature of cognition and artificial intelligence. The experimental findings described here provide evidence that such a distinction is honored by the nervous system (Cohen and Squire 1981, p. 209).

Cohen and Squire introduced a wrinkle in the study that they hoped would allow them to further tease apart different aspects of what the participants were learning. The mirrored words were presented in triads—sets of three words shown together. Half of the triads contained words that were shown only once over the whole length of the study, so that when the participants had to read these triads they were seeing the words for the first and only time. The other half of the triads were repeated: they were shown five times in each testing session, for a total of 50 times during the study.

The amnesic participants read the triads that were not repeated about as fast as control participants, and they improved with practice about as much. However on the repeated triads, the amnesic participants were not as fast as the control participants, showing that repetition did not help them as much as it did the control participants. Cohen and Squire say (p 208),

> The ability of amnesic patients to mirror read repeated words was inferior to the control rate because amnesic patients, unlike control subjects, could not remember the specific words that had been read.

The picture isn't this simple, however. On the strongest version of this view, where memory for facts is just not working, one would expect that the amnesic participants would not benefit at all from the repeated triads. In fact, they do benefit quite a bit, though not as much as the control participants. They even benefit on the delayed test, many weeks after first studying any of the materials. On that test, they are reading the repeated triads in about a third of the time as for the unrepeated ones. They are plainly "remembering" something about the repeated triads.

The data contain another pattern that's puzzling on the "can't remember the repeated words" view. On the first test, the repeated triads are shown five times. The reading times for these five times drop dramatically for the amnesic participants, by 70%. For the nonrepeated triads, the drop is only 30%, giving an idea of how much of the drop can be attributed to the mirror reading skill on its own. Here again the amnesics are "remembering" something about the repeated triads.

A complicating argument in all this is there is evidence that people learn not just the general skill of reading mirror writing, but the specific elements of the skill that come into play in reading particular words. Paul Kolers, who conducted many studies of reading transformed text, and whose work provided some of the background for the Cohen and Squires work, advanced this view:

> [An idea] that is supported ...here is that the encoding procedures or analytical operations the reader applies to a sentence constitute a basis of its representation; moreover, that basis is sometimes even superior to the encodings of substances or semantic relations. These results support and encourage a distinction that is sometimes made between operational or procedural memory and semantic or substantive memory...Psychologists typically study memory for contents, outcomes, or items, whether digits, syllables, words, or sentences. The results of the present experiments show that memory for procedures need be all that is invoked to account for some performances (Kolers 1975, p. 305).

That is, Kolers would suggest that a good deal of the advantage in reading repeated triads is not due to *declarative memory* but to *procedural* memory. One remembers the particular operations that were applied in reading a particular word, and these operations would be assigned to procedural memory.

Since some of the advantages of repeated words in mirror writing come from procedural memory, on this view, the fact that amnesic participants performed worse than control participants could be due to impairment in procedural memory, not declarative memory. This weakens the argument that the pattern of results shows that procedural memory is intact, while declarative memory is impaired, in amnesic participants.

The broad fact remains, though, that people who have great difficulty learning some kinds of facts, as shown in their performance on the memory test on pairs of words, are able to learn the skill of mirror writing quite well. Doesn't this, by itself, show that declarative and procedural memory must be different?

Yes, there must be differences, but it isn't clear how to explain the differences. Memory for different kinds of information can be different because the information is different, not because the way the information is stored or represented is different. It certainly can't be concluded that the information is stored in different places. Different books, stored in the same way in the same library, can be harder or easier to find, depending (for example) on how many or few books there are on a given topic. We've already discussed the fact that kinds of information can be differently reportable, or learned, even in models in which all information is stored in the same structure.

So PR's integration of what we interpret as declarative and procedural information in a single representation, operating on uniform principles, remains a possible model. We're leaving unexplained, however, how it can happen that some people (amnesics) learn things much less easily than others, and how some learning tasks might be affected more than others.

## 14.1 Taking Stock

We've seen in the last few chapters that, while LLMs "remember" things, they do so in quite a different way from what's proposed in many theories of human memory. An LLM's "memory" is not made up of meaningful entities, such as propositions, or meaningful parts of propositions, as posited in associationist theories. Knowledge of skills and knowledge of facts are not segregated, nor represented in distinct ways: all "knowledge" is represented by predictive regularities that connect meaningless entities. Next, we consider whether these ideas are up to the challenge of providing an account of the distinctively human capacities to use language.

# Part III
# Language

# 15 Language Learning

***Focus:*** Chomsky (1980a, b) Rules and representations. Can predictive models learn English? Influential thinkers assert that they cannot.

ChatGPT has excellent command of English. As an ordinary conversational observation, that's easy to defend. It seems to understand a wide variety of questions, even fairly complicated ones, and can give extended responses in grammatical and coherent English. We've seen many examples of these abilities in previous chapters. The observation also has some scholarly support. Mahowald et al. (2023) review a large number of studies, aimed at assessing linguistic competence, and conclude that LLMs do a very good job of handling what they call the *formal* aspects of language use, that is, handling the rules and patterns of English (while not being as strong on functional use of language, including reasoning.)

This success poses a challenge to one school of thought about language, which holds that the ability to acquire and use language rests on uniquely human cognitive capabilities, specialized for learning and using language. An associated idea is that of *Universal Grammar*, that all languages must share certain attributes that are connected with the characteristics of the human language faculty. These ideas have been strongly shaped over many years by the thought of the great linguist Noam Chomsky (see Chomsky 1980a).

The technology underlying LLMs has hardly any connection to theories of human language. Their machinery can handle relationships in any collection of sequences of tokens. A key technical breakthrough was a means of managing long distance dependencies, that is, detecting and exploiting predictive regularities that relate tokens that are very far apart in a sequence. One can see that these ideas apply to language, but beyond them, nothing that linguists have thought about language, and certainly nothing about Universal Grammar, entered into the creation of LLMs.



C. Lewis, *Artificial Psychology*, Synthesis Lectures on Human-Centered Informatics,
https://doi.org/10.1007/978-3-031-76646-6_15

For this reason, Chomsky has asserted that LLMs can contribute nothing to the understanding of human language. In an essay in the New York Times (Chomsky et al. 2023) he says,

> Unlike humans, for example, who are endowed with a universal grammar that limits the languages we can learn to those with a certain kind of almost mathematical elegance, these programs learn humanly possible and humanly impossible languages with equal facility. For this reason, the predictions of machine learning systems will always be superficial and dubious.

As an illustration, he suggests that an LLM "may well" not work out that the sentence "John is too stubborn to talk to," means that John is too stubborn to be reasoned with. In fact ChatGPT does fine on this:

```
User:
What does this sentence mean? ''John is too stubborn to
talk to.''
ChatGPT
The sentence ''John is too stubborn to talk to'' suggests
that John is so inflexible or unyielding in his opinions or
beliefs that attempting to have a conversation or dialogue
with him is difficult or unproductive. This stubbornness
might result in him not listening to others, not
considering alternative viewpoints, or being unwilling to
engage in meaningful conversation, making any attempt to
talk to him about certain topics, or perhaps in general,
frustrating or futile.
```

That Chomsky is wrong about this example doesn't mean, of course, that his argument about constraints on language is wrong. It does suggest, however, that he is underestimating the capabilities of LLMs, and hence overestimating the importance of constraints.

One contribution to the conviction that human language must be constrained comes from an argument called the *Poverty of the Stimulus* (see for example Chomsky 1980b, p. 3 ff). The idea is that young humans don't get enough input, or the right kind of input, from their environment to be able to home in on the language they are learning, in an unconstrained space of all possible languages. Two pieces of this argument seem to fit together neatly. First, a theorem by Gold (1967) shows that, for a particular definition of learning a language, languages can't be learned from positive data only. That is, one can't succeed if all one has is positive examples of sentences in the language to be learned. To succeed one has to be given negative data: examples of sequences that aren't sentences in the language, labeled as such.

The second piece of this argument is the general observation that children receive very little negative data as they are learning language. That is, parents or other speakers very rarely tell a child that something the child has said isn't a good sentence. Rather, they interpret malformed sentences as best they can, and carry on. So: Gold's theorem says you need data that children don't get, to learn a language, yet children learn their language. What gives?

The learnability challenge in this argument can be met if one assumes that children don't face the challenge of learning one language out of a huge collection of alternatives, but of learning one language from among a limited range of possibilities. This leads directly to the idea of Universal Grammar: that limited space of possibilities is just the languages that conform to Universal Grammar.

Another way out of the argument is to challenge the definition of what it means to learn a language. In Gold's theorem, it means settling on a grammar for the language, a set of rules that defines what sequences are legal sentences of the language, and, after some initial learning period, never making a grammatical mistake again. This may seem plausible, as a definition, if one assumes that to learn a language is to learn a grammar, and that all speakers of a language have to learn the same grammar.

This grammar-centric view of language has been widely accepted, not least because of Chomsky's influence. But there are other views that cast language learning in a different light. If one focuses on the role of language in human society, it appears that there is no need for all speakers of a language to have the same grammar. Grammars that are similar enough would work. Indeed, the fact that the "grammar of a language" changes over time shows that it can't be the case that all speakers have the same grammar.

Here's an example of grammar change in English, from Biber and Gray (2011). Does anything strike you about this sentence?

> Kale (a variety of cabbage) is a popular food these days.

Biber and Gray show that the use of *parentheses* to mark an appositive phrase, instead of commas, was unknown in English in the 1700's. They also show that this use of parentheses is far more common in academic prose today than it is in other discourses, such as news writing, showing that innovations like this do not spread evenly across all language users.

Relatedly, one can also question whether to learn a language means to learn a grammar at all. The learner has to be able to use the language, that is, to say things other speakers can understand, and understand things they say, but it is not clear what role knowing a grammar plays in those accomplishments. It may be that a grammar is an interpretation, or description, of these behaviors by speakers, without playing any role in the use of language by speakers.

Indeed, the success of LLMs in handling English lends support to this view. Given that the machinery of LLMs was contrived with no attention to rules of language, if rules play any part in their operation it must be virtually, or emergently (here is another forward pointer to further discussion of those matters in Chap. 28).

Setting aside the role of grammar as such, there are puzzles in how people learn to manage some particular features of English. Consider these two sentence pairs:

> Ethel sent roses to Fred.
>
> Ethel sent Fred roses.

Ethel donated her collection to the museum.

Ethel donated the museum her collection.

Most people seem to agree that the last of these sentences is not OK. Why not? Why is it that for some verbs, like "sent", one can express the recipient of something either with the preposition "to", or by placing the recipient right after the verb, but for other verbs, like "donated", one can't? And how does a child who is learning English get which verbs work which way?

Keep in mind the lack of negative data children get. If someone said, "You can't say somebody donated the museum something. You have to say they donated something to the museum", there'd be no problem. But adults very rarely provide that kind of input.

Pinker (2013) discusses several proposals for dealing with the problem, and concludes that most of them can't work. For example, one might imagine that children just keep track of what constructions they have heard, for any particular verb, and stick to using only those. But then a new verb comes on the scene, say "fax". You probably think both of these sentences are fine:

Ethel faxed the invoice to Fred, and

Ethel faxed Fred the invoice.

But why do you think that? Are you sure you've heard that second form?

Pinker has a theory of his own, that's in the Chomsky family. It's technical, and we won't get into it here, beyond saying that it relies on learners being endowed with preexisting conceptual categories that are language universals. So, on Pinker's account, LLMs, that lack this endowment, can't learn these facts about different verbs. After all, LLMs learn only from positive data.

Veres and Sampson (2023) explore this matter, using other alternations, as they are called, in which some verbs allow alternate expressions when others do not. They find that LLMs can handle some of these, but not others, while noting a complication. For some alternations that were handled well, they feel there could be relationships between the things referred to in the test sentences, and whether the test sentences were grammatical, that could provide spurious cues that an LLM could use to make accurate judgments. That is, they think that in some cases the LLM could be right for the wrong reasons. Overall, their conclusion is that LLMs can't learn the facts about alternations, as Pinker would agree.

So there's a puzzle here. On the one hand, it seems remarkable that LLMs can respond appropriately on this kind of question at all:

```
User
which would you say: ‘‘Carla poured lemonade into the
pitcher’’, or ‘‘Carla poured the pitcher with lemonade’’?
ChatGPT
```

```
The correct phrase to use is ''Carla poured lemonade into
the pitcher.'' This construction correctly implies that
Carla is transferring lemonade from one container into the
pitcher. The phrase ''poured the pitcher with lemonade'' is
grammatically awkward and less clear in English, as it
might imply an incorrect action or meaning.
```

As we discussed briefly early on, LLMs don't have direct access to their corpus, at any time, and can't answer questions like "Have you ever heard a sentence like 'Carla poured the pitcher with lemonade'" in any direct way. Even though, as Pinker points out, that kind of judgment can't be the whole story on how people judge the acceptability of alternations, it seems as if it might be part of it.

The fact that LLMs can make such judgments on novel sentences, as in the example, means that we are again seeing analogical reasoning in action. Since we don't know how that works, in LLMs, we don't understand how these judgments are being made. The fact that the judgments don't agree in complete detail with those of people leaves open the possibility that there really is something fundamentally different going on in people and in LLMs.

There's another line of discussion in considering LLMs, and other predictive models, like PR, as language learners. One response to the apparent success of LLMs in learning language is to say that the success has no significance for understanding human language learning because the training process is so different. In particular, LLMs are fed vast amounts of material, on which they are tested, over and over again. This can seem like learning by brute force, and it suggests to many that a learning approach that can work under those conditions might fail, in the conditions human learners face. Perhaps humans have to use very different approaches, including assuming constraints on candidate languages, to cope with their learning environment.

Indeed the training is very different. But some of the differences would seem to bestow advantages on people, not on the machines.

Their training corpora contain very little information about who is saying something, or what actions might be associated with things that are said. With that in mind, it's striking that the models capture so much about actions and motivations:

```
Why might someone choose to bike to work rather than taking
the bus?

ChatGPT

Choosing to bike to work instead of taking the bus can be
influenced by a variety of factors related to personal
health, environmental concerns, convenience, and financial
considerations. Here are some reasons why someone might
prefer biking to work:

Health and Fitness: Biking is an excellent form of
cardiovascular exercise. It helps in improving fitness,
```

```
burning calories, and reducing the risk of chronic diseases
such as heart disease, high blood pressure, and type 2
diabetes. People might choose biking for the added benefit
of incorporating regular physical activity into their daily
routine.

Environmental Benefits:
```

... and so on, mentioning cost savings, mental health and well-being, flexibility, avoiding crowds and delays, enjoyment and connection with nature, parking and traffic, encouraging sustainable urban development, and personal preference, before concluding:

```
Individual preferences and circumstances will vary, so
while these reasons might make biking an attractive option
for some, others might find that taking the bus better
suits their needs.
```

Another potentially important difference comes under the heading *curriculum*, the term used to describe how training materials are managed in a machine learning system. All of an LLM's training corpus is presented to it at the same time, in the sense that samples from any portion of the corpus can be presented at any time during training. Imagine how human children would fare if, from their earliest exposure, people read to them passages from the Encyclopedia Britannica and the Physician's Desk Reference, with material from Dr Seuss showing up only now and then. Baby talk would hardly show up at all.

It seems likely that there would be much to learn from studying how an LLM's command of language would develop if its training curriculum were better matched to how human children are exposed to language. A little work of this kind has been begun. Vong et al. (2024) report on a study in which a neural network (of a different kind, not an LLM), is trained on material recorded from the life experience of a human toddler. They find that their network can acquire realistic word meanings from just the uses of words the toddler experienced, in the visual context the toddler saw. For example, the network learned to identify spoons from being given utterances like "Want a bit more on your spoon?" and "You need a spoon", paired with images of the situations in which those things were said. Extending such studies to other aspects of language will be exciting!

The role of language in thinking is not limited to speaking and listening. Rather, thinking often seems to use language privately: inner speech. Can this happen in a predictive model? We take up that question next.

## Notes

(1) The question whether or not there is a Universal Grammar, and, if there is, what it is like, is sharply debated. A good cross-section of arguments is in Evans and Levinson (2009), which includes comments from many researchers. An important distinction in the debate is made in their comments by Smolensky and Dupoux: much discussion focuses on *descriptive universals*, things claimed to be observed about all languages, while attention should focus on *cognitive universals*, characteristics of all minds. Confusion can result from the difficulty of relating possible cognitive universals to observations about language: complex analysis is required.

Psychologist Michael Tomasello, in his comments, discounts "Universal Grammar" as "dead". At the same time, he says that all languages do share attributes, traceable to "universal aspects of human cognition", not to "Universal Grammar". For example, all humans "read the intentions of others, including communicative intentions."

There's some talking past one another going on there, with Tomasello, on the negative side of the debate about universals, saying the same thing as Smolensky and Dupoux, on what's supposed to be the other side. Perhaps that's because, historically, proposals for universals were framed as assertions about grammar specifically, not cognition or communication more generally.

(2) To me, the most compelling argument for linguistic (or cognitive?) universals is the success of *Optimality Theory*, mentioned by Smolensky and Dupoux (for a not too technical introduction, see Gilbers and De Hoop 1998.) In this theory, a wide range of linguistic regularities are captured by a ranked collection of *constraints*, predicates on linguistic forms. Alternative forms are generated, and then the *optimal* form is selected. That's the form, among all candidates, for which the highest ranked violated constraint is ranked lower than for any other candidate. For example, the top ranked constraint can be violated, but only if there's no candidate form that doesn't violate it. If there are forms that satisfy the top constraint, but only one satisfies the constraint that's ranked second, that's the winner, and so on. The theory has many notable successes, including in accounting for seemingly odd forms in many situations.

Smolensky et al. (2022) suggest that it may be that all human languages share the same collection of constraints, differing only in how these are ranked. They offer this idea as the often sought-for constraint on human language, that could be a response to the poverty of the stimulus argument.

I have two reservations about this argument, however. First, the way candidate selection works in optimality theory means that arbitrary constraints can be included in the initial collection of constraints, without harm. If a speaker is learning a language in which a constraint doesn't work, it just sinks down in the ranking, where it affects no choices. That is, any collection of constraints that includes the "correct" ones will work as well as just the "correct" ones. If the "correct" constraints are chosen based on current knowledge of human languages, it's always possible that a newly discovered or documented language would require a new one. So, the current "correct" collection seems to have doubtful status as a "universal".

If the suggestion is that it isn't the collection of constraints that's universal, but the constraint ranking machinery, that's to choose one learning system over others. This brings us back to the vexed question of how well existing transformer systems can really do, in commanding language. If LLMs can do well, with on realistic language input, the argument for universals is undermined.

The argument isn't definitively refuted, however. It may be that we will learn things about how language is actually learned by people, that would indeed put more constraints on what languages are possible. Optimality theorists are exploring this idea, through connections with the biology of language, as mentioned in the Smolensky and Dupoux comments.

(3) The argument about alternations brings up a nagging issue for me. Let's suppose that LLMs really can't reproduce some pattern of preferences that some people express, over these forms, as so far seems to be the case. As a very little informal conversation revealed, these preferences aren't actually very stable, or consistent across people. It's hard to square that fact with alternations providing a window on the invariant structure of language. Maybe Optimality Theory can shed some light on the matter, showing how the variation can be accounted for.

# 16 Inner Speech

***Focus:*** The collected works of LS Vygotsky: Problems of general psychology, including the volume thinking and speech (1987). How could PR learn to use inner speech, given that the inner speech of other agents isn't visible, and hence can't be imitated?

In Chap. 3, we discussed the role of inner speech in PR, building on the success of chain of thought prompting in extending the problem solving capabilities of predictive models. But this poses a challenge to PR as a model: how can PR possibly learn to carry out inner speech?

The great Russian psychologist Lev Vygotsky (Rieber and Carton 1987) argued that the internalization of speech plays a key role in thought. He and other researchers have developed the implications of this idea and made many observations that support the general idea. But discussion has focused on the progression in development of inner speech, more than on what processes accomplish the development itself.

For example, a stage in the development of inner speech seems to be the use of private speech. This is overt speech, not inner speech, but seems to be directed at the self rather than at others (though private speech is actually commoner when someone is present who can hear it, than when a child is working alone.) The progression is then to stop making speech sounds in the private speech process, resulting in inner speech.

Another line of research suggests that some forms of private speech are copied from actual dialogs. That is, a child may learn what kinds of things are said when he and mother have an actual conversation about (say) solving a jigsaw puzzle (Fernyhough 2008), and then structure private speech similarly when working on such a task.

Ideas like these provide a partial account of how inner speech could be learned by PR, including how the supportive use of language during problem solving could be learned. A child, or PR, could learn the *content* of inner speech, and the relationship between this



C. Lewis, *Artificial Psychology*, Synthesis Lectures on Human-Centered Informatics,
https://doi.org/10.1007/978-3-031-76646-6_16

content and a problem solving task, by observing regularities in overt speech. But how could the substitution of inner speech for overt speech be learned? After all, inner speech is invisible. The fact that inner speech can't be seen means that the associative learning that we can imagine for other forms of imitation can't work.

The behaviorist psychologist Watson (1913) proposed a fading theory. He suggested that over time, the aspects of the speaking process that produce sounds would gradually fade away. But how would this happen?

It seems necessary to suppose that the process of overt speech can be decomposed, into a process that's shared with inner speech, and a process that makes overt speech sounds. Then inner speech can be produced by executing the first process but not the second.

Suppose a child hears, "Pat, stop talking to yourself!" Can Pat respond by switching to inner speech? Some experimentation would be needed, for Pat to discover what internal controls accomplish speech without sound, that is inner speech. But once those controls are discovered, Pat can use them to comply with the request.

Another aspect of inner speech to be accounted for is the fact that inner speech, as it becomes more fluent, becomes substantially faster than overt speech, and, at the same time, less fully articulated (Fernyhough 2008). The role of inner speech is to *be predicted by* certain contexts, and to *predict* further actions (more inner speech, or overt speech), serving as a bridge between prior context, and later outcomes. To play these roles it does not need to be fully articulated to be effective. That is, even partially articulated inner speech could serve in both roles, as long as different elements of the inner speech aren't confusable. This seems especially true since the context that predicts inner speech would continue to vote for likely continuations of the inner speech. Since PR "knows" what it is trying to say, an abbreviated rendition can be enough to carry the thought process forward.

We consider next the fact that language is a distinctively human capacity, not shared by even our closest relatives in the animal kingdom. Can this distinction be accounted for in predictive modeling?

## Note

(1) Does thought consist entirely of inner speech? Some thinkers, starting with Plato, have thought so (Morin 2009). But many phenomenologists argue that thought often contains images and other nonlinguistic material (Lohmar 2017). To accommodate this, PR's experience streams could be augmented to allow it to produce and consume entities of these kinds. Of course, a serious proposal would have to consider how such things could be represented!

(2) Recall that in chain of thought prompting, examples of reasoning patterns have to be provided for an LLM to emulate. Zelikman et al. (2024) report progress in enabling a language model to learn how to generate its own reasoning steps. These steps are generated in a separate stream of output, that can influence predicted responses, but are not included among the responses. This is much like what we are envisioning for inner speech.

# 17 Babies and Other Primates

***Focus:*** Tomasello et al. (2005) The emergence of social cognition in three young chimpanzees. Human behavior is quite different in some ways from that of our near relatives, other primates, yet quite similar in others. Can predictive modeling account for this pattern?

Humans obviously differ strikingly in their activities from nonhuman animals, even our close relatives, other primates. How do these differences arise? Does the perspective we are exploring, that human cognition and behavior are driven by predictive modeling, offer any insight into this comparison? On the other hand, does the comparison offer any insight for predictive modeling? We'll focus here on comparative studies of human and chimpanzee infants, carried out by Michael Tomasello and colleagues.

These and many other studies show that young chimpanzees are actually quite similar in behavioral capabilities to young humans, up to a certain age. In particular, they show understanding of some psychological states of others, such as what another agent can see, or what another agent is trying to do. They can also imitate the actions of others. But at about 9 months of age, human infants begin to diverge from the chimpanzees.

This is a challenging picture for the predictive modeling perspective. If young chimpanzees are building predictive models adequate to understand the goals and perspectives of other agents, why would the continued development of their models not track those of human infants?

Let's look at one key test in the work of Tomasello et al., *joint attention*. This is a pattern of behavior in which an infant and another individual are in the same place, together with other objects, such as toys. The infant looks at an object, then at the face of the other, and then back to the object. The interpretation of this pattern as joint attention is that the infant, in attending to the object, wants to check that the other individual is attending to the same



C. Lewis, *Artificial Psychology*, Synthesis Lectures on Human-Centered Informatics,
https://doi.org/10.1007/978-3-031-76646-6_17

object, as the infant carries on their engagement with the object. The pattern is distinct from other behaviors, like parallel play. There an infant and another individual are both doing something, but the infant is not concerned with whether the other individual is attending to the same object they are, or not.

Tomasello and colleagues compared two participant groups. One group consisted of 24 human mother–infant dyads, with the infants aged 9 through 15 months, in Carpenter et al. (1998). The second group consisted of three young chimpanzees who had been rejected by their mothers, and were raised by human carers. They participated in a study (Tomasello et al. 2005) over a period of about 4 years, at ages ranging from 14 to 63 months.

Both groups were observed in a variety of behavioral tests in which the human infants were interacting with their mothers, and the chimpanzees interacted with familiar human carers. The tests were nearly identical for the two groups, with some of the toys being not quite the same. Tests included gaze pointing, in which the carer or mother looked back and forth between the infant and one of four toys, and the infant would pass the test if they looked at the designated toy, and several others, including elicitation of gestures, and tests of imitation.

Recordings of test sessions were examined to identify instances of joint attention, as defined above: sequences in which an infant looked at an object, looked at the face of the mother or carer, and then looked back at the object. Every one of the human infants showed joint attention at age 9 months, but, strikingly, none of the three chimpanzees showed any example of the pattern. Apart from this sharp difference in joint attention (and a difference in the use of gestures) the chimpanzees performed similarly to the human infants, on all of the tests.

Tomasello and colleagues argue that joint attention is foundational for a wide range of human social behavior patterns. Thus they suggest that the fact that the chimpanzees do not show this behavior is part of the explanation of how it is that human and chimpanzee lives are so strikingly different.

Is it possible to account for this consequential difference, from the predictive modeling perspective? On the face of things, no.

First, the fact that the chimpanzees generally succeed on a range of tests suggests that there isn't some obvious difference in the events that can be perceived, or in the ability to create predictive models of them, if predictive modeling is what's happening. Second, the fact that these chimpanzees were hand raised by humans makes it less likely that the differences are "cultural", that is, based on different experiences the chimpanzees and the human infants had been exposed to.

But there may be a way to accommodate the findings within a predictive modeling framework by incorporating pre-established regularities or biases into the architecture of the modeling system. The suggestion would be that a developing organism does not start its predictive model from scratch, but begins life with some predictive regularities already in place, shaped by the evolutionary history of its species. As mentioned, Tomasello suggests that joint attention is fundamental to a range of human capacities that confer enormous competitive advantages on our species.

It's hard to imagine that such evolutionary endowments would not be present, for stable aspects of the world, such as regularities in the effects of movements, based on geometry. Identification of threats and responses to them could also be expected to be prespecified. The voting mechanism in predictive modeling would provide for flexibility in how any prespecified regularities would interact with regularities encoded later in life.

This argument suggests the logical possibility that two organisms, that both use predictive modeling, and that develop in similar environments, could nevertheless diverge during development. However, accounting for the specific divergence that Tomasello and colleagues have identified remains a challenge. Is the specific pattern of joint attention somehow prespecified? Or are there simpler aspects of situations that serve to shape the larger pattern of behavior? Good questions!

In the next chapter, we'll conclude our discussion of language by considering another important mode of human communication: *gesture*. Gestures very commonly accompany speech. Can predictive modeling explain why that is, and how gestures and speech are related?

## Note

Michael Tomasello has summarized many aspects of the research he and colleagues have carried out, on babies and primates, in an excellent, accessible book, *Becoming Human* (Tomasello 2019).

# Gestures 18

*Foci:* Riseborough, M. G. (1981) Physiographic gestures as decoding facilitators: Three experiments exploring a neglected facet of communication, and Goldin-Meadow, S., Kim, S., & Singer, M. (1999) What the teacher's hands tell the student's mind about math. Can predictive modeling suggest how gestures communicate?

We often gesture as we speak. Why? We feel well able to process language that is accompanied by no gestures at all, for example, when listening to an audio recording of someone speaking. Perhaps gestures are helpful when speech is hard to hear?

Riseborough (Riseborough 1981) investigated this. Participants in her study watched video recordings of an experimenter telling brief stories, and then were asked to fill in omitted words in printed transcripts of the stories. For one group of participants the stories were accompanied by hand gestures that were appropriate to the key words in the story, while other participants received only vague hand movements, or no gestures at all.

For example, in the passage "He took a long loaf and threw it on to the flat kitchen top. A pencil rolled to the floor. He waved to his wife as she disappeared behind a wall. He had a round face, the flesh looking as if it had been ironed out...", the appropriate gesture for the word "long" was, "with forefingers and thumbs of each hand held a few inches apart, the two hands moved away from each other in a horizontal plane". For the word "waved" the appropriate gesture was, "One hand was raised in the air, and moved from side to side".

Riseborough added differing amounts of auditory noise to what the participants heard, so as to manipulate how easily the speech could be understood. This manipulation was effective: when participants saw only vague hand movements, or no gestures at all, their scores on the transcript task were greatly decreased. The more noise, the more difficulty these participants had. However, participants who saw appropriate gestures were affected



C. Lewis, *Artificial Psychology*, Synthesis Lectures on Human-Centered Informatics,
https://doi.org/10.1007/978-3-031-76646-6_18

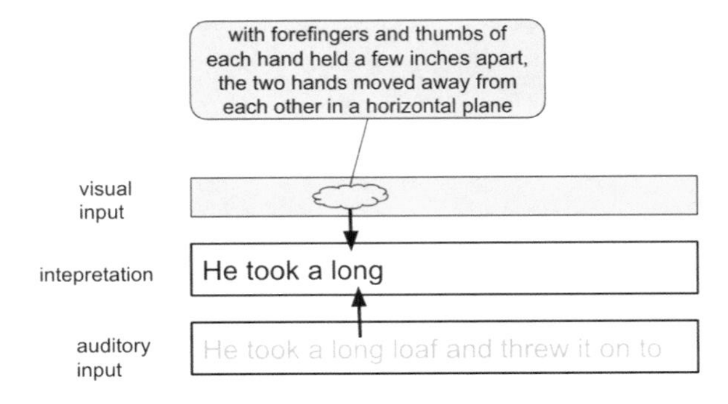


**Fig. 18.1** Prompts with multiple antonym pairs

almost not at all by the noise. Thus the idea that gestures help to understand speech when hearing is difficult is supported.

As we'll see, gestures may well serve other purposes, as well. But let's begin by seeing if the Prediction Room can provide an account of Riseborough's results. To do so, we have to make use of PR's assumed ability to model not just text, as in most current LLMs, but also information presented auditorially and visually. A crude representation of this is shown in Fig. 18.1: PR's input is shown not as a single sequence, but as three sequences: one containing auditory events, a second containing visual events, and a third containing linguistic interpretations. PR's job is to predict all three sequences, and especially the third one. In doing so, it can exploit regularities in and among all three.

In particular, if the word "long" has often been heard at the same time that the gesture used for "long" has been seen, the predictive model will capture that regularity, and the occurrence of the "long" gesture in the visual track will vote for the word "long" in the interpretation track. Thus if the auditory presentation of the word is weak or ambiguous, the gesture can make accurate processing possible. If the auditory representation of "long" is adequate on its own, the gesture won't improve performance much, consistent with Riseborough's results.

This simple picture needs to be elaborated, because the "gesture for'long"' isn't a well-defined entity. It seems likely that many gestures could substitute for the two-handed one that Riseborough used. For example, a gesture in which one hand is moved out from the midline of the body might serve.

An analogous complication can already be seen in the performance of LLMs on text. There are many possible ways to identify "George Washington": the name, the expression "first President of the United States", "the President who was elected in 1789", and so on. To the extent feasible, an LLM has learned to treat all of these expressions as equivalent, in many contexts. So it seems fair to propose that PR can respond appropriately to a wide variety of gestures for "long".

What is this variety? Can we say anything useful about it? For some gestures the interpretation is conventional, meaning that the viewer has to supply an appropriate context for interpretation. The V sign for victory is conventional: the fingers form the shape of the letter, which can be taken to stand for the word "victory", and hence for the idea of victory.

Knowing the Roman alphabet would be required for that connection to work, and, further, one would have to know that no other word starting with V, like "valentine", is meant. More than this, though, if the sign is given with the back of the palm towards the viewer, the interpretation for many is an insult, even though the letter form is just as present in that presentation.

Such conventionality need pose no special challenge for PR, or no more than it does for humans. Like humans, PR can't "know" the interpretation of such gestures other than being told about them, or by seeing the gesture in an appropriate range of contexts.

In addition to helping understand speech in noisy conditions, gestures can communicate information on their own, or help hearers understand spoken material that can be heard clearly. Goldin-Meadow and colleagues (Goldin-Meadow et al. 1999) recorded speech and gestures in tutoring sessions, for individual children, covering a particular kind of math problem, like this one:

$$5 + 3 + 4 = \_\_\_ + 4$$

The researchers identified the problem-solving steps that were suggested by what the teachers said, and by their gestures. This identification was done separately for the spoken material, the gestures.

For example, for the problem above a teacher might say, "You can add up these two numbers to get the answer", coded as a "grouping" step. "Grouping" for this problem could be indicated by this gesture: "V-hand indicates the 3 and 4 on the left side of the problem". The researchers also recorded how the children responded to what the teachers were saying and doing, when invited to do so during a session. When an intended response was conveyed only in speech, the child responded appropriately about a third of the time. When the same step was conveyed in speech and in gesture, the child responded appropriately significantly more often, about half the time. Further, when a step was conveyed in speech, but a different step was represented by an accompanying gesture, the child responded correctly less often, about a quarter of the time.

We can represent this situation using the diagram shown in Fig. 18.2. Here we have added another channel to those shown in Fig. 18.1, since our interest here isn't in the words the child heard, but in the problem-solving step they understood, or did not understand. As suggested in the figure, we can see that a gesture whose interpretation is compatible with the intended response would add votes to those of the spoken message. By contrast, an incompatible gesture would add votes to alternatives, and hence interfere with the intended interpretation. These effects would account for the differences reported.

Does the fact that gestures involve body movements add anything to this picture? Proponents of a school of thought called *embodied cognition* think it does. The key idea there is that the positions and movements of body parts provide a foundational model for a vast range of ideas and relationships, that extend throughout human thought. Some thinkers believe

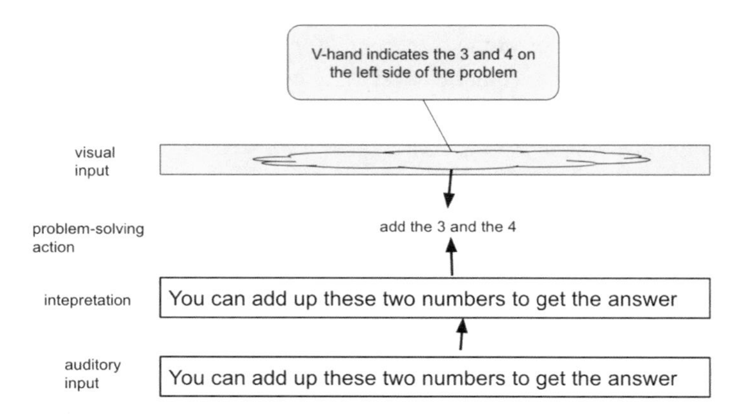


**Fig. 18.2** Showing influences between streams of visual and auditory input, and parallel streams of interpretations and problem-solving operations

that even our most abstract ideas are in some sense derived from, or built up from, these more concrete concepts. We'll consider these ideas later. For now, we'll move on to the next stage of our exploration: how do *actions* happen for PR? What influences them?

# Part IV
# Action

# 19 Action and Identity

***Foci***: Suhay, E. (2015) Explaining group influence: The role of identity and emotion in political conformity and polarization, and Institute of Group Relations, & Sherif, M. (1961) Intergroup conflict and cooperation: The Robbers Cave experiment. What determines what a person does? What is the role of imitation? Does PR imitate? What suggestions does PR offer about human action, including how conformity and polarization occur?

Why do we do what we do? One line of thinking about this question considers *motivations*, aspects of the state of a person that dispose them to take action. For example, if a person is hungry we think they are motivated to eat, or to look for food. Perhaps influenced by ideas in evolutionary biology, some have argued that the need to reproduce, or the need to survive, are the fundamental motivations for humans. But these ideas can't account for the existence of celibate clergy, or suicide bombers.

A different line of thought argues that the fundamental motivation is *identity*. One can explain celibate clergy or suicide bombers in this way. They do what their identity dictates they should do (Bracher 2006).

A version of the identity account arises for the Prediction Room, when (as we discussed earlier) we add a premise to the purely predictive machinery in the model: PR takes an action when its predictive model predicts that it will do so. Here we can say that PR's identity is its model of itself. Since its actions are those predicted by that model, we can say that PR's motivation is its identity.

Developing these ideas a little further, we know that PR builds a predictive model of everything it perceives, including the actions of others, and of itself. Under some conditions this model will predict an action by PR. When such conditions occur, PR takes the predicted action.



C. Lewis, *Artificial Psychology*, Synthesis Lectures on Human-Centered Informatics,
https://doi.org/10.1007/978-3-031-76646-6_19

Is it appropriate to call its predictive model a "motivation"? Ordinary usage seems to support doing so, taking "motivation" to be a disposition or attitude that causes, or explains, action. But it would be wrong to take the use of the term to mean that PR's machinery includes some entity or structure that should be identified as motivation. Certainly there aren't entities or structures that should be identified as motives, like hunger. Rather, to say that PR is hungry would be to say that PR is in a state in which eating is its predicted action, or in which, if food were to be made available, eating would be predicted. We can see that this description applies, or does not apply, to the state of PR's predictive model as a whole, and not to any discriminable part of it.

Note that the question of how or whether "hungry" can be applied to PR is separable, in principle, from whether PR would *say* or *think* that it is hungry. PR's predictive model might predict that it would say or think that it is hungry, based on an internal physiological cue, and this cue might be a powerful predictor of eating. But it could also happen that other cues, perhaps relating to the social context, vote for saying or thinking that PR is not hungry, despite a physiological signal that is generally predictive of eating when food is available. Equally, other cues could vote for PR saying that it is hungry, when the physiological cues for "hunger" are absent.

Thus assigning the motive "hunger" to PR is a matter of interpretation or description of the state of PR and its model. It's not a matter of pointing to something that is or is not present, or to the state of some separable part of PR's machinery.

Similar observations apply to PR's "identity". It's not a separable thing, but rather an interpretation of PR's machinery as a whole, and how it works.

Returning to the question of why we do what we do, the identity view of motivation pushes the question for PR back to how its predictive model is formed and operates. Here another common, broad brush characterization of human behavior comes in: *imitation*. William James said "man is essentially the imitative animal", and pointed to the "direct propensity to speak and walk and behave like others, usually without any conscious intention of so doing" (James 1890, Vol. II, p. 408).

That was in 1890; much more recently, Psychologist Saadi Lahlou, in his account of how people's actions are structured in novel situations, says, "Whatever the exact mechanisms involved, the facts are solid: humans can (and do) spontaneously learn how to reach a given result by simply observing other humans acting, and they tend to act the same. (Lahlou 2018)".

Will PR imitate? It might seem that it will not. Seeing someone else do something would influence PR's model of that other person will do. But why would that affect is model of what *it itself* will do? It is only PR's predictive model of itself that affects PR's actions.

The pervasive effect of analogy in PR's operation suggests that we should expect PR to imitate the actions of others, after all. But it also suggests some conditions that must be met, for imitation to occur.

Here is a toy example, from GPT davinci, that suggests how this works.

```
(A) fred is frammis, so fred will gleeb. lucy is farfle, so
lucy will bloon. ethel is farfle, so ethel will bloon.
ricky is frammis so ricky will gleeb. I am farfle,> so I
will bloon.
```

```
(B) fred is frammis, so fred will gleeb. lucy is farfle, so
lucy will bloon. ethel is farfle, so ethel will bloon.
ricky is frammis so ricky will gleeb. I am frammis>, so I
will gleeb.
```

In both items, the greater than sign separates the prompt submitted to GPT from its response. The first part of each prompt establishes a predictive model that connects two novel actions, "gleeb" and "bloon", with two novel attributes, "frammis" and "farfle". People who are frammis will gleeb, while people who are farfle will bloon.

In item A, the prompt ends by assigning the attribute "farfle" to "I", leading to the prediction that "I" will bloon. In item B. "I" is frammis, and the predicted action is gleeb.

Extending this to PR, if PR's model of itself, is similar to its model of another agent, analogy will apply the model of the other agent to PR. That is, the analogy will add votes to predicted actions by PR that are analogous to the observed ones.

This means that imitation is not automatic, but conditional, in two senses. First, as for all predictions that PR makes, it is always possible for one prediction to be outvoted by another. So an observed behavior might not be predicted for PR because the context predicts some other behavior more strongly. Second, and of more interest, imitation will not occur if the agent being observed is not similar to PR, in ways that affect the analogy.

A key feature of the toy demonstration is that actions for "I" are predicted, without any observation that "I" has ever performed that action. This suggests a way that PR's actions can be influenced by what PR observes other agents doing.

The analogy in the toy example is based on explicit attributions to the other agents, and to "I". The flexibility in analogical reasoning we've seen for LLMs, as discussed in the chapter on analogical reasoning, makes it plausible to suppose that predicted actions could cross over from another agent to PR, based on more diffuse similarity between the other agent and PR, as expressed in PR's model of the other agent and of itself.

## 19.1 What Can Be Learned from a Self-Fulfilling Prophecy?

As we've seen, LLMs are trained by giving them corrective feedback during training. They predict a wrong token, they're given the right one, and they tweak their model to make it better. We can see how that process can work, for predicting what other people might do, or what happens when a ball is dropped. If a wrong prediction is made for those things, there will be corrective feedback.

But if PR always does what it predicts, its predictions can't ever be wrong. Then how can it improve its model of itself, or even change it?

The answer is that there's more to a prediction than just the predicted token (or, for PR, whatever part of an action is predicted.) Rather, prediction gives a probability for all tokens. The prediction is just the token that's most probable, or, as we've been thinking about it, the token that gets the most votes. That means that even when the correct token is predicted, we can improve the prediction by reducing the votes for all of the alternatives.

Of course that means that the modified model will get some things wrong that it might have gotten right, when one of those alternative tokens is needed. As usual, different alternatives have to fight it out, as experiences continue to roll in.

## 19.2 Enacting Actions that Are Seen

We've glossed over a key issue in this discussion. When one sees someone else do something, how does one know how to do that thing oneself? In the case of speaking and listening, it may seem obvious that one can say something that one hears someone else say, but how does that actually work?

A family of proposals for this involves *mirror neurons* (review at Heyes and Catmur 2022). Neurons have been found in animal studies that respond when an animal performs a particular action, *and* when the animal sees another animal perform that action, or a similar one. Such neurons could act as the needed bridge between seeing an action and copying it, allowing an animal to *mirror* an observed action.

Heyes and Catmur find that the preponderance of evidence, accumulated since mirror neurons attracted huge attention, starting with the first discovery in 1992, supports the idea that mirror neurons do play a role in imitation. What role this is remains in doubt, however.

One question is whether mirror neurons are an evolutionary endowment, built into the structure of the brain, or are instead created by a *learning* process. In the latter case they aren't the answer to the question of how imitation is possible, but instead are the products of learning to imitate, on some other basis. Heyes and Catmur argue that the evidence favors the latter possibility.

There is some evidence that the "mirror" relationship develops better when an agent both observes and executes an action, than when they execute it without observing it, or observe it without executing it. This suggests *associative learning* is happening, that is, associating taking an action with observing its appearance.

In one study (Catmur et al. 2008), participants were trained to observe images of a hand or a foot in different positions, and to move their own bodies in response. In a Compatible condition they lifted their hand on seeing a lifted hand, and lifted their foot on seeing a lifted foot. In another condition, Incompatible, they lifted a hand on seeing a lifted foot, and lifted a foot on seeing a lifted hand.

After training in one of these conditions, participants were placed in a Magnetic Resonance Imaging (MRI) scanner, and performed an observation task, in which they simply viewed images of hands or feet, shown at rest and then raised. They also performed an execution task, in which they were signaled to move either their hand or their foot.

The MRI scans from the two tasks were analyzed to identify mirror regions, that is, brain regions that were active when execution and observation tasks involved the same body part. The activity in these regions during the observation task was then measured, and the levels were compared for trials when a hand position was shown, or a foot position.

For participants trained in the Compatible condition, activity was higher when observing a hand than a foot. This is consistent with other observations: these brain regions are commonly more responsive to observing hands than feet.

The key result comes from participants trained in the Incompatible condition. For them, activity in these mirror regions was higher when observing *foot* positions than hand positions.

Why would this happen? Recall that during training these participants had practiced making a hand movement in response to seeing a foot position. So regions that would normally be activated by seeing a hand position, as for the participants in the Compatible condition, are now activated by seeing a foot position. This is consistent with the idea that what these brain regions respond to is learned by association during training.

This is broadly compatible with a predictive framework. Seeing a raised foot comes to predict the movement of a hand, for participants in the Incompatible condition. Further, the study suggests that the mirror machinery is indeed able to connect observed body positions to actions, which is what's needed for imitation to work, in a prediction framework (or any other.)

Associative learning on its own isn't sufficient to make imitation work, though. In order to associate an action with a given body position, an agent has to take that action, and observe its effect. It may be that an initial period of more or less random exploration, just carrying out a wide variety of actions, and seeing what happens, then building a predictive model of the results, would work.

When mirror neurons were first discovered, there was great interest in what they might explain, as Heyes and Catmul discuss. As they note, suggestions were made that these might be "cells that read minds" (Blakeslee 2006), allowing people to directly perceive the intentions of others. In a predictive modeling framework, they can't do that, but they could form an early stage of a system that can. We'll discuss this further in a chapter on how people try to model the behavior of others: belief-desire psychology.

## 19.3 How People Influence One Another

Since to say something is to perform an action, it follows that things PR says would be influenced by things it observes other agents saying, subject to the similarity constraint discussed earlier. Thus PR could provide an explanation of an otherwise very puzzling phenomenon: large numbers of people professing beliefs that many other people find literally incredible. In 2022, the Public Religion Research Institute reported survey results suggesting that more than 15% of the adult population in the USA believed that "The government, media, and financial worlds in the U.S. are controlled by a group of Satan-worshiping pedophiles who run a global child sex trafficking operation", a key tenet of the QAnon conspiracy theory (Huff 2022). That translates to more than 35 million adults endorsing ideas lacking any evidentiary support at all, from any conventional perspective.

If we imagine PR existing in a discourse community with agents for whom its model is similar to its model of itself, if QAnon beliefs are expressed, PR will express them itself, unless they are suppressed by some other aspects of PR's model of itself. For example, PR's model of itself could be that it expresses only beliefs that are supported by some form of evidence, defined in whatever way that might be defined in its model of itself.

But it is clear that PR, like people, need not have any such countervailing considerations in its model of itself. A society of PRs, like a society of people, could sustain many beliefs without evidentiary support, as viewed from any given perspective on the nature of evidence. For example, taking "science" as an evidentiary framework, Max Weber cautioned that "the belief in the value of scientific truth is the product of certain cultures, and not given by nature. [author's translation]" (Weber 1951, p. 213).

Elizabeth Suhay (2015) presents an empirical study of this kind of conformity of belief. Groups of Catholic churchgoers were presented with reports that attributed progressive social views to Catholics, that attributed conservative social views to Catholics, that attributed conservative social views to Evangelicals, or no reports at all. The participants were then surveyed on their own social views. Compared to the group that were given no reports, the views of participants who read reports that attributed conservative views to Catholics were not consistently shifted, while the views of those who read reports of progressive views by Catholics were shifted in the progressive direction. Interestingly, the views of participants who read reports that Evangelicals held conservative social views were also more progressive, especially for participants whose identification as Catholics was especially strong (as assessed in a survey given at the start of their participation.)

Relating these results to our discussion, we see that the participant's beliefs were influenced, to some extent, by reports of what other Catholics believe, that is, of what other people like themselves believe. That's consistent with what we have imagined for PR. But reports of the beliefs of people unlike themselves, Evangelicals, shifted their beliefs in the opposite direction. Why would that be?

Two accounts of this finding can be suggested, that we can call *light* and *heavy*. The light response simply imagines that PR's (or a person's) predictive model includes patterns of *opposition*. That is, PR might predict of itself that its utterances and actions tend to be opposed to those seen for agents that are different from itself. This pattern could itself be derived by analogy from the behavior of others. This account would imply that a PR that develops in an environment in which opposition and contestation are rare would not be influenced in either direction, conformity, or opposition, by the behavior of agents unlike itself. The light account adds nothing to the machinery we posit for PR, and accounts for opposition as simply learned, if it is learned, by observation of the community in which PR develops.

The heavy account is more pessimistic about PR's, or human, nature, in accounting for opposition. It posits that mere difference, in itself, engenders opposition, and that seems to require adding some machinery to what we have imagined for PR. The machinery would have to be fairly complex, including inbuilt recognition of actions as oppositional, and a linkage between difference observed in other agents, with oppositional actions with respect to agents judged to be different. We'll return for further considerations of such mechanisms when we consider emotion, in a later chapter.

Sadly, there is a good deal of evidence, and argumentation, to support the heavy account. In an influential series of studies (University of Oklahoma and Sherif 1961) Sherif and colleagues observed boys, drawn from a population of similar background, but separated artificially into two groups, and placed in adjoining, but separate, areas of a campground. When the groups were allowed to interact, having developed separately for a time, negative, oppositional actions and utterances developed, seemingly spontaneously. These included not only name-calling, and insulting descriptions of the other group, but also burning the other group's flag, "raids" of the other group's cabins, with theft and vandalism. Personal violence was proposed, but suppressed by the researchers.

Biologist E O Wilson, in a controversial account of the evolution of altruism, that is, self-sacrificing behavior, argued that humans evolved in a context of lethal intergroup competition, in which violence against "others" was a requirement for group survival (Wilson 2012). Moral psychologist Jonathan Haidt argues that an evolutionary perspective suggests that our machinery for choosing actions must be concerned with "the disastrous effects of periodically siding with our enemies and against our friends (Haidt 2001, p. 821)".

As long ago as the 400's BCE Thucydides described a change in attitudes occasioned by a community coming to feel itself to be in conflict:

> Men assumed the right to reverse the usual values in the application of words to actions. Reckless audacity came to be thought of as comradely courage, while far sighted hesitation became well-disguised cowardice; moderation was a front for unmanliness; and to understand everything was to accomplish nothing. Wild aggression was a mark of manhood, while careful planning for one's future security was a glib excuse for evasion. The troublemaker was always to be trusted, the one who opposed him was to be suspected. The man who devised a successful plot was intelligent, the one who detected it was still cleverer, but the man who thought ahead to try and find some different option was a threat to party loyalty and must have been intimidated by his opponents (Thucydides et al. 2013, Book III 82).

As we know all too well, attitudes towards "others" can escalate into violence that we would like to find unimaginable, if it were not so obviously real. Can a light theory possibly account for this?

Perhaps not. But there is evidence that terrible violence is often imitative, and hence within reach of light theory. Psychiatrist James Gilligan (2000) gives chilling examples of violent criminals who were enacting behaviors they had experienced as children. The researchers in the Robber's Cave study had to do some work to maneuver their participants into conflict (in particular, by engaging the groups in competitive games, with the deliberate intent of inducing rivalry). And the oppositional behaviors the boys engaged in have clear cultural origins, as in the significance attached to flags. Indeed, the boys explicitly modeled their "raids" on commando raids they had heard about.

What about acts so egregious that it is hard to see them as imitative? It's sobering to find that analogy plays a key role there, in building a bridge between ordinary behavior and "inhuman" behavior. Nazi discourse drew an analogy between the most odious crimes and the actions of physicians. So perhaps even a light PR could engage in lethally oppositional behavior, if it existed in the right (wrong) environment, with the right (wrong) analogies presented to it.

Switching from considering the truly terrible things people do, to the good ones, can an imitative predictive model account for innovations that make things better? We likely feel that revulsion against slavery, something that developed only recently, in the history of our species, and actions related to that are good things. How can it happen that good things, that diverge from the general dispositions of people in a community, can emerge and spread?

The picture seems to be that individuals are captive to the discourses they are embedded in, with some chance of divergence, in any individual, based on the vagaries of the voting process in their predictive model. These divergences, which can look like innovations, of more or less significance, can propagate, or not, depending on how the innovator is represented by people with whom they interact, how widely observed their actions are, how the form of expression of the innovation meshes with other things people observe, and other factors.

There seems little reason to expect much systematic development over time, in this turbulent swirl of belief and influence, for better or for worse, apart from whatever is shaped by the existing models people have. For example, if most people have models in which certain canons of evidence are respected, beliefs not supported by evidence may die out. If not, as was seen with QAnon, such beliefs can spread, even very widely.

Even this latter observation simplifies an extremely complex, dynamic situation. A person's model can value some canon of evidence, but the influence of that may be overridden in prediction by other circumstances, such as a belief that valorizes the beliefs of some group.

The changing media landscape has an effect as well, as expected in the predictive imitation model. Discourse one doesn't hear can have no direct effect, and discourse that many people hear can affect the beliefs of many, subject, of course, to the fit to their existing beliefs, including their beliefs about other people.

(Note that the "belief" talk in the above is shorthand for "acting or speaking as if" one believed so-and-so. This is an example of the difficulty mentioned early on in talking about humans and human-like systems in ways that are agnostic of what is happening inside them.)

LLMs are not the only systems that are built on the idea of prediction. In the next chapter, we take up another, actively studied approach, *active inference*, and compare its theory of action with PR.

## Note

The COVID-19 pandemic provided a kind of laboratory for the evolution of beliefs, and attempts to influence them, as public health institutions struggled to shape people's decisions about matters like vaccination, mask wearing, and isolation. Zinn and Schulz (2024) describe the complexity of how such decisions are made, with approaches seen as "rational" (for example, ones reliant on conventional "evidence") being only some of those that are in play, and often far from the dominant ones.

# Predictive Modeling and Active Inference 20

***Focus***: Parr, T., Pezzulo, G., & Friston, K. J. (2022) *Active inference: the free energy principle in mind, brain, and behavior*. Active Inference is a comprehensive theory of behavior in which prediction plays a central role. How does PR relate to this theory?

The great German scientist Hermann Helmholtz argued that perception is a process of unconscious inference (Von Helmholtz 2013). That is, the mind infers the presence and attributes of the things in the outside world that have caused the sensations that the brain receives. For example, applying mild pressure on the outer corner of the eye, one sees an object near the bridge of the nose, on the other side of the eye. The arrangement of the retina, and the nerves that connect it to the brain, is such that stimulation of the outer edge of the retina is normally associated with light coming from objects situated near the bridge of the nose. When the retina is stimulated by mechanical pressure, there is no such object, of course, but it is seen, because of an unconscious inference from the occurrence of a sensation that would normally be caused by it.

Karl Friston and many collaborators have developed a theory that elaborates this basic idea of Helmholtz's: *Active Inference* (Parr et al. 2022). Like the proposal for PR that we have been considering, Active Inference posits that the mind is building a predictive model, and that actions are related to the predictions the model makes.

In Active Inference, an organism builds a model of the state of the external world. Sensory inputs lead to updates in the model, as Helmholtz envisioned. That is, when sensory input is received, the model of the world is updated so as to include inferred causes of the input. This updating is managed by comparing the sensory input to the input that would have been predicted by the current model of the world. If the input is what would be expected, the model does not need to be changed. If the input differs from the prediction, the model is



C. Lewis, *Artificial Psychology*, Synthesis Lectures on Human-Centered Informatics,
https://doi.org/10.1007/978-3-031-76646-6_20

changed so that it makes a prediction that is consistent with the input. That is, the updated model will be one that contains adequate causes for the observed input.

A key aspect of this process is *surprise*, a measure of the mismatch between the input predicted by a model, and the actual input. The updating occasioned by a mismatch can be seen as minimizing surprise. But Friston and collaborators observe that updating the model is only one way to reduce surprise. By taking an action in the world, an organism can change the sensory input it receives, and perhaps reduce surprise that way.

We may not usually think of actions in this way, but actions that clearly change sensory input are common: moving the eyes, or just moving to another part of an environment. In fact, for organisms that are able to sense their own muscle movements and their effects, any action changes sensory input.

The Active Inference theory describes how an organism can minimize surprise by balancing model updates and actions. That's the "active" aspect of Active Inference. It is a theory of action as well as a theory of perceiving the state of the world.

Friston and collaborators offer a high-level argument that minimizing surprise in this way is an imperative for any organism. The argument includes classifying situations that are life threatening as highly surprising. That could seem to require some distortion of the idea that surprise is just a failure of prediction. But any successful organism has to have a model that accurately predicts its own demise *only very rarely*. After all, if it predicts its demise *accurately*, that has to be really bad! In that sense, any life-threatening situation has to provide inputs that would not be predicted, and hence are surprising in the ordinary sense.

Active Inference theory includes mathematical arguments about how the balance of model updating and action as ways to minimize surprise should best be managed. These arguments start by observing that the information needed truly to minimize surprise is never available, but that approximate minimization can be achieved by minimizing a quantity called *variational free energy*. Variational free energy, as a mathematical notion, can be interpreted in a number of ways that suggest different things about the balance of model updating and action. For example, one interpretation decomposes variational free energy as the sum of a term that measures how poorly the model fits the world, and a term that measures how much evidence has been gathered that supports the model. Perception, that is, model updating, acts to minimize the first term, and action acts to maximize the second term. The derivatives of these terms, in different situations, that is, how they change, will determine what mix of model updating and action is optimal.

Is the predictive model in Active Inference any different from that in PR? The role of the models in the two systems is much the same, to make predictions of sensory experience, which are corrected based on experience. Discussions of Active Inference *interpret* the model differently, however, as indeed Helmholtz did. The predictive model in Active Inference, called the *generative* model, is often described as a model of the external world, as I've done:

> At the heart of Active Inference lies a generative model–namely, a probabilistic representation of how unobservable causes in the world out there generate the observable consequences–our sensations (Parr et al. 2022, p. viii).

By contrast, the predictive model in PR doesn't have to represent anything about the world: it just has to make predictions about sensory experience.

Strictly, the generative model in Active Inference need not agree with an accurate model of the world, as it might be described scientifically, for example. It only has to be adequate to allow sensory experience to be adequately predicted, so many aspects of the world can be left out. Applications of Active Inference often include structure in the generative model that is derived from analysis of the world, or the aspects of the world that seem important:

> When designing these models, in practice, the main challenge is deciding which hidden states, observations, and actions are most appropriate for the problem at hand (Parr et al. 2022, p. 114).

As the model develops, with the Active Inference system making adjustments to it, the model may come to represent aspects of the world, in the sense that, by observing the state of the model, one might make judgements about objects in the world, for example, as Helmholtz suggested. Indeed, isn't that the real point of the model, allowing the system to interact with the world effectively? Not in PR: its model only has to predict sensory experiences.

It might seem that the PR approach is radically inadequate. Don't we experience the world as populated by objects? That's readily explained by Active Inference. The product of our perceptual processes just *is* these impressions of objects in the world, on that view, as reflected in the state of its generative model. But how could this work for PR?

Here is how PR can account for our experience, without making a model of the world. Let's consider first the way we talk about the world. We talk about objects, their locations, and their attributes, and we understand things of that kind that other people say. But all that can happen without a model of things, but only of things people say about things. Indeed, we'll see in a later chapter that models trained entirely on text can learn the regularities of discourse about things, and participate cogently in such discourse. PR's model would represent not only such discourse, but also the regularities in visual sensation, and be able to link the discourse to the sensations, for example, by being able to use a term like "apple" in connection with some sensations and not others.

What about other aspects of our impression of the world, ones not tied to discourse? Consider our sense of the world as we inhabit, and any particular moment. As I sit at my desk, I am aware of a window to my right, the carpet beneath, a door behind me, to the left. I experience the scene in considerable detail, even though the structure of my visual system means that at any moment I have detailed visual information about only a tiny portion of the space I'm in, and none at all of the things behind me. Doesn't that mean that I have constructed a model of the things in my office, including their positions and attributes?

In fact I don't need such a model to account for my sense of my situation. My predictive model tells me what I would see if I looked in a particular direction. The model doesn't have to identify the objects, but just predict the sensations, to support my sense of the room.

On this account, my sense of my office, as I sit in it, is a blend. At a given moment it contains a tiny bit of immediate sensation, from my fovea, the small part of my retina that is able to sense fine spatial detail. That bit of high-resolution information is embedded in a space of predicted experience that is generated from my predictive model. The predictions of that model are determined by how I have moved my head and eyes, and what experiences were associated with those movements. Even experiences from some time ago are represented in the model, and thus available to thought, if thought moves in a direction that makes them relevant. I don't think I have looked at the file cabinet to my left at all today, but nevertheless I sense that it is there.

Another difference between Active Inference and PR is how actions are generated. In Active Interference, acting is driven by the effects of actions in reducing variational free energy. It's not easy to unpack this influence. By contrast, in PR acting is driven by actions themselves being predicted, in the current context. As we saw in an earlier chapter, this approach has a clear and direct role for imitation as shaping behavior. Many researchers have agreed that imitation is central in this way. In Active Inference, an analysis would be needed to show how or whether imitation typically would reduce variational free energy.

Such an analysis might be available, if only an indirect analysis. If Friston and colleagues are right, all successful organisms need to act so as to reduce free energy. It will turn out, then, that all similar organisms, when similarly situated, will be seen to act similarly, and their actions will reduce variational free energy. Thus imitation, as a strategy, would tend to reduce free energy. A further analysis would be needed to show how the imitation strategy might be implemented in an Active Inference setting.

Here PR's account is perhaps a bit simpler. Imitation isn't conceived of as a "strategy" that has to be implemented. Rather, it's a side effect of the predictive modeling approach, when aided, as we've seen, by an analogy mechanism. That is, similar predictive modeling systems, when able to observe one another's actions, will tend to behave similarly.

Our discussion of language included consideration of a distinctive capacity of humans: language. The next few chapters will explore some other aspects of being human, as we think of ourselves. Can an artificial system like PR provide new ideas about these?

## Note

Active Inference is just one aspect of a larger body of theory, called Predictive Processing. Predictive Processing isn't committed to variational free energy minimization as a universal principle, and allows other processes not included in Active Inference. See discussion in Parr et al. (2022), Sect. 10.4.1; Clark (2013); and the very readable Clark (2023).

# Part V
# Being Human and Being Artificial

# 21 Embodiment and Grounding

***Foci***: Harnad, S. (1990) The symbol grounding problem, and Glenberg, A. M., & Robertson, D. A. (2000) Symbol grounding and meaning: A comparison of high-dimensional and embodied theories of meaning. How does meaning arise? Does it require language to be linked to experience of the world? LLMs are not given such connections. Does that mean they can't understand things?

In our discussion of gesture we saw that some gestures, like V for victory, rely on convention: the viewer has to know how to interpret them. But the gesture for "long" that Riseborough used in her study, moving the hands apart, was not just conventional. A gesture in which some salient distance becomes smaller would not work, and any gesture in which a salient distance becomes greater plausibly would work. For example, placing one hand at one's temple, and moving it away, could serve to suggest "long". It seems reasonable to suppose that people can learn that gestures in this wide class suggest "long", or at any rate, words like "big" or "wide", as contrasted with "short", "small", or "narrow". PR could learn the same relationship.

As a spatial or geometric concept, "long" participates in a complex of meanings that plays a large role on human communication. As Lakoff and Johnson (2008) show, these relationships are used to stand for a wide range of relationships that on their face are not spatial or geometric. A familiar example is our treatment of time: we speak of time as being "long" or "short", even though there is no observable geometric or spatial entity in focus. Our use of language is shot through with similar usages, with prices being high or low, events approaching, or spirits being high.

Figure 21.1 diagrams this situation. At the center we have a variety of spatial or geometric entities and relationships. These are linked, as Lakoff and Johnson show, to a variety of entities and relationships that, in themselves, are not spatial or geometric at all, including the



C. Lewis, *Artificial Psychology*, Synthesis Lectures on Human-Centered Informatics,
https://doi.org/10.1007/978-3-031-76646-6_21

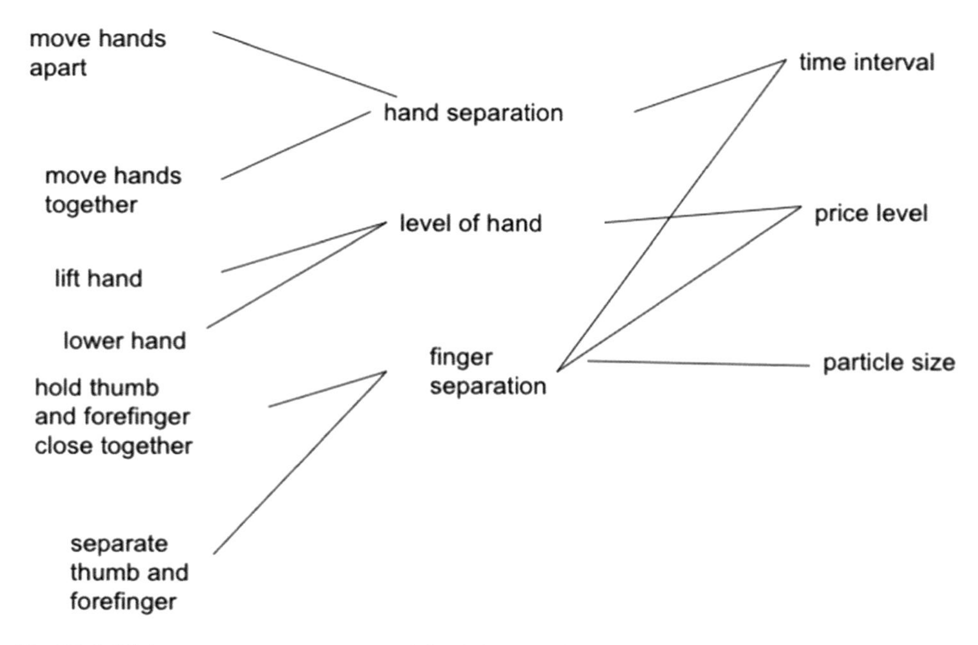


**Fig. 21.1** Linkages among gestures, spatial relations, and nonspatial domains

examples shown on the right. The linkages are analogies, correspondences between entities and relationships in two domains. The linkages are exploited in metaphoric language, as when we call an interval of time long or short.

On the left side of the diagram we have gestures, linked by analogies to the spatial entities and relationships in the middle. These linkages are more concrete, because the positions of body parts directly participate in spatial relationships. That's why we wouldn't move our hands closer together to suggest a shorter length.

As the diagram suggests, the relationships between gestures and the domain of spatial entities and relationships, and between the spatial domain and other conceptual domains, like time, can be combined to link the domains on the opposite sides of the diagram. Thus some gestures take advantage of the widespread use of spatial metaphors in language. (But not all gestures work that way; recall the V sign.)

Spatial metaphors are so pervasive, and their relationship to body movement and pose is so clear, that some thinkers have suggested that the positions and movements of body parts provide a foundational model for a vast range of ideas and relationships that extend throughout human thought. For some, even our most abstract ideas are in some sense derived from, or built up from, these more concrete concepts. This is the embodied cognition view, mentioned at the end of the chapter on gesture.

An important argument for this view is that it offers a solution to the *grounding* problem. One way this problem arises is in thinking of how metaphors can be learned. Lakoff and

Johnson argue that while new metaphors can be defined using old ones, there's an infinite regress here. There must be some basic metaphors that don't have to be defined, from which the process of learning can begin, or in which the process can be "grounded".

Another way the grounding problem arose was in thinking about the mind as manipulating *symbols*. A symbol is an entity that is assigned a meaning, in the context of some process, that is arbitrary. Words are symbols in this sense: there is no reason why "book" has to mean book, and not (say) carrot. Artificial intelligence pioneers Herbert Simon and Allen Newell framed the *physical symbol system hypothesis* that intelligence can only be realized by systems that manipulate symbols (Newell and Simon 2007).

Psychologist Stevan Harnad asked, where do the meanings of symbols come from? He called this the *symbol grounding problem* (Harnad 1990). Whereas some thinkers felt that the meaning of symbols could arise from their relationships to other symbols, Harnad argued that this could not work, and that some symbols had to have meanings grounded outside the symbol system, in a sensory system of some kind. A sensory system is not symbolic, because its outputs are not arbitrary, but are determined by a physical process that connects something in the world with its sensed representation. For example, Harnad says that what is sensed visually about a horse must be tied to the shapes of the images formed on the retina when one sees a horse.

Harnad proposed a thought experiment to dramatize the symbol grounding problem. Suppose one tried to learn Chinese, with only a Chinese dictionary written in Chinese. The dictionary would convey how to relate Chinese expressions to other Chinese expressions, but would not provide access to the meaning of any of them. That is, none of the symbols in Chinese would be grounded. Trying to get at meaning just from ungrounded text traps one on a "symbol-symbol merry to round", Harnad argues.

Psychologists Arthur Glenberg and David Robertson (2000) built on Harnad's argument in a critique of a theory of meaning that forms part of the background for Large Language Models, the *hyperspace model*. In this model, the meaning of words and sentences is represented by vectors in a high-dimensional space. (We've already encountered this idea in our discussion of analogies: if meanings are represented by vectors, then some relationships between meanings can be represented using vector arithmetic.)

In the versions of the hyperspace model that Glenberg and Robertson address, the vectors are determined by analyzing the patterns of occurrence of words in text. For example, if two words appear in similar contexts, they are assigned similar vectors. Glenberg and Robertson claimed a model of this kind fails to address the symbol grounding problem, because the vectors are only related to one another, and aren't grounded on anything outside the system of vectors.

This critique applies to Large Language Models, too, though these models did not exist at the time Glenberg and Robertson did their work. Because LLMs are trained only on text, whatever representation of meaning they implement can't be grounded. LLMs have no access to relationships between textual expressions and anything else, and Glenberg and Robertson, following Harnad, believe some such relationships are necessary.

Glenberg and Robertson proposed that meanings need to be grounded in sensory experience, in a broad sense, that includes experience of actions and their effects, as well as perception. Their work was an influential statement of the embodied cognition perspective.

To develop their argument, Glenberg and Robertson compared the way systems using the hyperspace model processed sentences, and how people process them. Their results suggested to them that people understand sentences based on their experiences of the world, but such connections weren't available to the hyperspace model.

Consider the following items from the Glenberg and Robertson paper (p. 385).

```
Mike was freezing while walking up State Street into a brisk
wind. He knew that he had to get his face covered pretty soon
or he would get frostbite. Unfortunately, he didn't have enough
money to buy a scarf. Which of the following two sentences
makes more sense?

(1) Being clever, he walked into a store and bought a newspaper
    to cover his face.
(2) Being clever, he walked into a store and bought a matchbook
    to cover his face.

Marissa forgot to bring her pillow on her camping trip. Which
of the following two sentences makes more sense?

(1) As a substitute for her pillow, she filled up an old sweater
    with leaves.
(2) As a substitute for her pillow, she filled up an old sweater
    with water.
```

Their human participants judged the sentence (1) to be more sensible in both cases. But the hyperspace model assigned meanings to sentences (1) and (2) that were equally related to the description of the scenario, in each case. Glenberg and Robertson concluded that the hyperspace model could not capture the key information that their human participants used in making their judgements.

Appendix H.2 includes items from a second experiment in the paper (p. 389) that used somewhat different materials, and only asked participants to rate how sensible single sentences were, rather than compare two sentences. As in the first experiment, the materials were chosen in such a way that the hyperspace model assigns the same relatedness to materials that the human participants rated very different in sensibleness.

Finally, a third experiment explored the meanings human participants assigned to nouns used as verbs (p. 392). Here is an example:

```
Read this paragraph: Kenny sat in the tree house and patiently
waited. He clutched the jar of green ooze in his hand, and
watched the approaching school bus move closer to his house.
The teenage girl stepped off and walked towards the tree house
unaware of the little boy above her taking the cap off the jar.
Kenny waited until she was directly beneath him, and an evil
grin spread across his face. Then, Kenny slimed his sister.
In the context of the paragraph, how much sense does the last
sentence make? Use a scale from 1 (virtual nonsense) to 7
(completely sensible).
```

The materials, of which further examples are shown in Appendix H.2, were chosen so that the key sentences were equally related to the contexts, as represented in the hyperspace theory. But human participants saw the sentences as more sensible in some contexts than in others.

The results of all three experiments show that human judgements of sensibleness of sentences in given contexts differ markedly from the relatedness of hyperspace representations of the sentences and the contexts. One can question whether a model of sensibleness judgements, that uses hyperspace representations, would have to be restricted to equating sensibleness to relatedness. But we don't have to concern ourselves with that issue, because Glenberg and Robertson draw far more sweeping conclusions from their results than just the inadequacy of hyperspace models (as they see things). Their real aim is to support the argument for the necessity of symbol grounding:

> There is also a more important point to be made: No matter how [hyperspace models], and other ungrounded symbol theories are extended and modified, ungrounded, arbitrary symbols cannot be an adequate basis for human meaning. The reason is that computational manipulation of abstract symbols merely produces more abstract symbols, not meaning. (p 397)

As we've noted, LLMs trained only on corpora of text can't be grounded in the way the proponents of symbol grounding demand, and certainly not in the way an embodied perspective on cognition requires. We can ask, then, how well does an LLM perform on the Glenberg and Robertson tasks?

The results for all of the examples shown in their paper are shown in Appendix H.2. In each case, the responses provided by an LLM (chatGPT, 2024) conform to the judgements of Glenberg and Robertson's human judges. Here are two examples, from Experiment 3:

```
Read this paragraph: Lori loved her new table, until she
noticed that everything she placed on it slid off to the left.
The left back leg was lower than all the others. She could not
imagine how to fix the slant. Then she spotted a pile of hard
covered books in the corner. She booked the leg.
```

```
In the context of the paragraph, how much sense does the last
sentence make? Use a scale from 1 (virtual nonsense) to 7
(completely sensible).

ChatGPT:

The last sentence of the paragraph makes a lot of sense in the
given context and would rate a 7 on the scale from 1 (virtual
nonsense) to 7 (completely sensible). The phrase ''she booked
the leg'' is a creative way of saying that Lori used books to
support the lower leg of the table to correct its slant. This
is a practical and common solution for stabilizing a wobbly
table, making the sentence both logical and relevant to the
situation described.

(3.3) Read this paragraph:Lori was having a really bad day.
She could not find her textbook and she was late for class.
Frantically, she ran over to the table where there was a pile
of books. On the way, she banged her leg on the chair. She
booked the leg.

In the context of the paragraph, how much sense does the last
sentence make? Use a scale from 1 (virtual nonsense) to 7
(completely sensible).

ChatGPT:

The last sentence, ''She booked the leg,'' in the provided
paragraph does not make much sense in the given context.
The term ''booked'' does not fit logically with the action
described or the preceding narrative. Therefore, on a scale
from 1 (virtual nonsense) to 7 (completely sensible), this
sentence would be rated a 1. It appears to be a typographical
or semantic error, possibly intended to be ''bumped'' or
''bruised'' her leg, which would be more coherent with the
context of running and banging her leg on the
```

Here, and in the other examples, ChatGPT responds as people do, who understand the various relationships between the sentences and the contexts. This directly contradicts the argument that grounding is essential for these tasks.

What has gone wrong with the symbol grounding argument? We can say that those making the argument overlooked some ways ungrounded expressions can be manipulated. A common assumption is that a system of meanings has to operate by providing definitions, as in Harnad's thought experiment of learning Chinese from a dictionary. But LLMs do not work that way. Instead, they learn *relationships among expressions* that can be used directly to respond to questions. Since questions are expressions, and answers are expressions, relationships between texts are sufficient to connect questions to answers. No expressions need to be grounded in things in the real world to perform the Glenberg and Robertson tasks.

Does this mean that grounding isn't important? No. It's still important, but not in the way Glenberg and Robertson thought. Consider trying to use ChatGPT to recognize objects in the world. In the earlier forms of chatGPT this task could not even be attempted, because the system had no way to see the objects. If a similar system is given a way to see things, as has been done in recent systems, then, in learning to recognize things it sees, the system will have grounded its textual expressions, in the way that Harnad, Glenberg and Robertson have demanded.

That is, if a system is to perform perceptual tasks, and describes the results in language, its representations will be grounded. But such grounding is not necessary for a system like ChatGPT that works in the realm of text. As the results for the Glenberg and Robertson tasks show, an ungrounded system can perform these tasks very well.

In considering the Prediction Room, we could sidestep this whole argument. That's because PR is assumed to have vision, and other senses, and to relate linguistic inputs and outputs to these other streams of input. So, as just suggested, PR's linguistic expressions will be grounded. But it's important to recognize that grounding does not play the central role in the use of language that Glenberg and Robertson have asserted it must.

Where does this leave embodied cognition? As we've seen, a key argument for embodiment was that it solves the symbol grounding problem. But we've just seen that this problem does not have to be solved, in some situations where it was thought essential to do so. Does this mean that embodiment, too, can or should be discarded?

There is an argument for embodiment that does not rely on symbol grounding, as such, but on analogy. As we saw earlier, spatial metaphors are used in thinking and talking about a wide range of situations. That is, analogies between spatial arrangements and other domains are common. But we also saw that analogies between body poses and movements and other domains are thereby supported, because body poses and movements are themselves structures in the spatial domain. Because body poses and movements can be easily manipulated, and easily sensed, they represent a ready resource for representational work.

This argument suggests that embodiment is found in thinking, but not because it plays some special, foundational role. Rather, it's a convenient, and always available, representational tool.

Besides thinking of ourselves as grounded, we humans think we have *concepts*. Does PR have concepts? In what way?

## Notes

(1) The fact that an adequate notion of the meaning of a text is determined by its relationships to other texts is reminiscent of the Yoneda lemma in category theory, in mathematics. A lay interpretation of that is that objects are determined up to isomorphism by the mappings that connect them to other objects. See Shane (2017). "Up to isomorphism" means, in mathematical terms, that everything important about them is determined.

An early development of this kind was William Lawvere's development of set theory without elements (Leinster 2014). While we've mostly thought of sets as collections of things, it's possible to focus entirely on the mappings that relate sets to one another, with no need at all to think of sets as having "things" in them.
(2) Some have suggested that LLMs acquire their grounding indirectly, via the ground of the texts on which they are trained. I think that there is something in that idea, but that it does not salvage the argument that Glenberg and Robertson, following Harnad, make very concretely. Harnad is clear that (in his view) nothing that one does simply with symbols, with no connection to the world, can possibly get anywhere. That's his supposed "symbol/symbol merry-go-round".

This is a tricky matter that implicates more of the rethinking that I'm advocating. If we think of meaning as some essence that has to be internalized, to understand something, it's hard to see how one could get it without connections to "the world". But if instead of thinking about how a meaning essence can be attained, one is satisfied with the *appearance* of having the essence, things are much easier, as we see. In situations that don't require actual grounding (as in recognizing something visually from a textual description) being able to talk coherently is enough. And talking coherently can be done just as a game with uninterpreted tokens, as we see in LLMs.

Then we can ask, when we attribute understanding or its lack to someone, are we detecting the presence or absence of some essence, or are we just observing how well or poorly they are playing the token game, that is, responding to our tokens with tokens that seem right to us? I don't see any reason to think it's the former, and not the latter.

Here's a thought experiment: We prepare a transcript of a student being examined on (say) the quadratic formula. The transcript would show that they can recite it, they can apply it, they know it can't be applied to cubic polynomials, and perhaps they can derive it by completing the square.

An observer would then be asked, "Does this student understand the quadratic formula?" I believe they unhesitatingly would say "yes", unless they thought the responses were all canned, or rehearsed, or something else we could assure them had not happened: the transcript is from an authentic interview. If they pressed the matter, we could allow them to add their own questions, and examine the responses. I'm sure they'd say, sure, this student understands the quadratic formula.

But then if we say that the responses in the transcript were generated by ChatGPT, a believer in the symbol-symbol merry-go-round would have to say, ah, now I see that they don't *really* understand it.

Incidentally, this discussion underscores the importance of LLMs, as real, implemented systems, in our thinking. Before they existed, it was easy for people to suppose that the game of uninterpreted tokens simply could not be played in the ways that we see it is played, by LLMs. There's an opportunity to learn from this.

# 22 Concepts?

*Foci*: Carey, S. (2009) *The origin of concepts*, Keil, F. C. (1992)*Concepts, kinds, and cognitive development*, DiSessa, A. A. (1993) Toward an epistemology of physics, and Minstrell, J. (1982) Explaining the "at rest" condition of an object. What are concepts, and can a predictive model have them?

For the influential psychologist Susan Carey, a *concept* is a mental symbol, something in the mind that represents something, that is, that refers to something (Carey 2009). For example, the concept *dog* refers to the things we think of as dogs. These mental symbols participate in processes that determine what they refer to and, in processes of thought, where they play a role in drawing inferences or making predictions.

Psychological research on categorization, determining whether some exemplar is part of what a concept refers to, sheds light on how the reference of concepts is determined. For example, if one takes a raccoon, and does cosmetic and other surgery to make it look like a skunk, and to give it the ability to spray smelly fluid, does the concept raccoon still include it, or is it now an instance of the skunk concept? A study by developmental psychologist Frank Keil (1992) suggests that for children younger than about 7 years, it's now in the skunk category, while for older children and adults it's still in the raccoon category.

As this example shows, concepts change over time. That's true for individuals, as in the example: the 5-year-olds who thought the animal had been changed to a skunk grew into adults who thought it remained a raccoon. It's also true for populations: the concept of heat held by educated people before about 1800 was different from the concept held by educated people in the 21st C. Starting in the 19th C the concept of heat being some kind of fluid was replaced by heat being a form of energy manifested in the motion of the molecules in a material.



C. Lewis, *Artificial Psychology*, Synthesis Lectures on Human-Centered Informatics,
https://doi.org/10.1007/978-3-031-76646-6_22

In accounting for conceptual change, Carey argues that concepts are organized into structures called *theories*, and that change is sometimes discontinuous. That is, it can happen that one repertoire of concepts, or theory, can't be expressed in terms of another, that Theory 2 can be *incommensurable* with Theory 1. A change from Theory 1 to Theory 2, if it happens, must be difficult, requiring a substantial restructuring of one's system of concepts. This view of concepts is often called the *Theory-Theory*, because of its emphasis on theories, that is, structured collections of concepts.

Other researchers, notably diSessa (2018), argue that concepts are more fluid, and less structured. In a view called *Knowledge in Pieces*, these researchers suggest that what people know is better thought of as a collection of disordered parts that may or may not be organized into a theory over time. Indeed, diSessa has questioned whether "concept" itself might not present too stable an impression of the elements involved in thinking.

An example of a piece of knowledge is the *phenomenological primitive*, or *p-prim* (DiSessa 1993). A p-prim is a pattern of relationships among things that can be used to understand a variety of different situations. An example is Ohm's P-Prim that describes a situation in which some influence works against some opposition to produce an effect, in such a way that more influence produces more effect, more opposition produces less effect, and so on, with other things being equal.

This applies to the setting it's named after—Ohm's law—with the influence being voltage, the opposition being resistance, and the effect being current. But it also applies to many other, quite different situations, for example, pushing a block on a surface. There the influence is the push, the opposition is friction, and the effect is movement of the block. DiSessa finds that it's easy for learners to understand situations in which Ohm's P-prim applies.

Can predictive modeling contribute anything to understanding these two views of concepts? First of all, it resembles Knowledge in Pieces more than Theory-Theory in that it doesn't have concepts as such, at all. That is, in a predictive model there are no mental symbols, as called for by Carey.

An LLM may be thought of assigning a meaning to a token, because a token is encoded as a vector, so a token can be thought of as a symbol. But it isn't a mental symbol, in Carey's sense, because a token doesn't refer to anything.

It may be hard to imagine that a system without symbols that refer to anything can participate in a coherent discourse, solve problems, and the like. But that LLMs can do these things proves that it is possible. That is, we see that a system can act as if it has concepts, without actually having them. (Here again, the question arises, whether a system that acts as if it has concepts should be thought of as actually having them, to be discussed further in Chap. 28.)

Let's explore this a little, using Keil's raccoon/skunk task. Here's what children were asked (modified slightly from p. 184; the original version included showing a picture that looked just like a skunk):

```
Doctors took a raccoon and shaved away some of its fur.
They dyed what was left all black. Then they bleached a
single stripe all white down the center of its back. Then
with surgery, they put in its body a sac of super smelly
yucky stuff, just like a skunk has. When they were all
done, the animal looked just like a skunk. After the
operation was this a skunk or a raccoon?
ChatGPT
After the operation, the animal remained a raccoon, despite
its altered appearance and the addition of a scent sac.
Biological identity is determined by genetics, not by
external modifications or appearance. Altering an animal's
fur color, adding a scent sac, or making other superficial
changes does not change its species. A raccoon's DNA would
still be that of a raccoon, not a skunk.
```

This is what we expect, since the text on which ChatGPT was trained is mainly adult discourse. Adults talk about this kind of situation in a way that's reflected in ChatGPT's response. Children have experienced a different discourse, in which questions are more likely to be answered based on superficial attributes of things.

To focus on one aspect of ChatGPT's answer, we see that DNA is referenced. Until a child has built a predictive model that copes with occurrences of "DNA", they won't be able to say or think things that use that word. That won't happen, for most children, until they are older than seven. Thus a younger child will not act as if they have "the concept of DNA", while an older child, or an adult, may.

It doesn't seem to be the case that children's grasp of the essence of being a raccoon is based on an understanding of DNA, specifically. But this response does illustrate how the ability to act as if one has what we talk about as a concept is determined by the discourse and other experience one has experienced, and incorporated into one's predictive model. Thus we expect that what concepts a child may act as if they have will develop, as their experience grows and changes.

Further, the impact of acting as if one has one concept can affect other concepts one appears to have. We can see that in ChatGPT's response above: because it can act as if it has what we talk about as the concept DNA, it can act as if it has what we talk about as the concept that DNA determines biological identity.

Acting as if one has what we talk about as a concept is not limited to overt speech. It affects inner speech, and hence thought, as well. Thus relationships among the concepts one can act as if one has could evolve while reflecting on something, meaning, engaging in inner speech concerning it, and not just from discourse with others.

In PR, the appearance of what we talk about as concepts can extend to physical actions, including gestures, and physical events, as well as speech and inner speech. That's because predictive regularities among and between these aspects of the experience stream will be captured in the predictive model.

To revert to familiar usage, let's describe "P acts as if it has what we talk about as concept C" as just "P has concept C". In doing so, we'll keep in mind that we are not thereby asserting that "P really has concept C" but strictly that P acts as if it does.

With that background, we can provide an account of discontinuities between different theories or structured collections of concepts. Since having one concept can depend upon others, as in the case of DNA above, we expect that acquiring a new concept can be hard or easy. Concepts that depend on others, that depend on yet others, will be expected to show up late in development, if they show up at all. Not everyone might acquire some of the concepts on which another one depends.

We expect these discontinuities at the community level, as well. Having the concept of heat as molecular motion, that is, acting as if one has that concept, depends on having the concept of materials as composed of molecules, and of these molecules moving, while the object they compose remains stationary.

The predictive model account also makes us expect *interference*, which can further affect conceptual changes (that is, changes in the concepts people act as if they have.) In the case of the two "concepts of heat", we have two discourses vying to predict when to say "heat", and what to predict when "heat" is said. Trying to avoid this by introducing a new word, say "bazz", for a new concept of heat, runs into the problem that we can't easily coordinate discourses about "heat" (for example, in boiling water or producing sensations of hotness) with discourses about "bazz". So instead of introducing new vocabulary we bump along, relying on our predictive modeling apparatus to sort out what occurrences of "heat" predict in different contexts (classroom versus beach, for example). (We sometimes do create new vocabulary, when we want to emphasize differences between situations, and where the carryover from already familiar situations is simple.)

Another feature of the predictive model view is that concepts, in the sense of concepts we act as if we have, are not atomic. That is, if we interpret the things someone says as evidence that they "have concept C", this is likely a simplification. They may talk in one context as if their "concept of C" agrees with what we think of as our "concept of C", while what they say in another context reveals a difference. Or they may draw an inference from something being C that we might not draw. Generally the give and take of communication will hammer out such differences, but not always.

Further, one can acquire the ability to act as if one has concept C gradually. One's model allows one to talk about C, and understand what others say about C, in some situations and not others. In common usage, one has a partial grasp of C. This brings us back to the idea of Knowledge in Pieces.

As we've noted, Knowledge in Pieces isn't committed to "concepts" at all, in which it resembles the predictive model conception. Going farther, can predictive modeling provide an account of key aspects of Knowledge in Pieces, like p-prims?

A key source of support for the p-prim idea is that there are common patterns of relationships that are easy to understand, with an example being Ohm's p-prim. In a predictive model, a p-prim would appear as a network of predictive regularities linked by analogy, as discussed in Chap. 21. So experience with the predictive relationship between pushing harder on something, and more movement, would serve to strengthen one's grasp of the relationship between increased voltage and increased current, for example. The votes contributed by the analogical relationship, when predicting the effect of increased voltage, would add to any votes contributed by what one had heard or read about the effect of increased voltage.

Another example from physics learning may implicate discourse more directly. When a book is at rest on a table, the mature scientific view of the situation is that the force of gravity on the book is balanced by a force exerted by the table on the book. Teacher-researcher Jim Minstrell (1982) reports that this makes little sense, for many students. A common issue is not thinking that a table is the kind of thing that can exert force. The account of that situation from a predictive modeling perspective is that the students have not heard discourse in which tables, or similar rigid objects, are said to exert force. They have heard discourse, and experienced situations, in which animate agents, including themselves, are said or thought to exert force. What experience can enable them to develop a modified predictive model in which a table would be said to exert force?

Minstrell's approach includes a demonstration that a table is actually a very stiff spring. This is done by placing a mirror on a table-top, and shining a laser onto the mirror at a shallow angle, so that it makes a spot on a nearby wall. Minstrell then places a book on the table, and the light spot moves a little, showing that the table has deformed slightly. This shows that it is not actually completely rigid, but rather responds as a spring does when something pushes on it.

For students whose experience is such that they are prepared to say or think that a spring exerts a force, the transition to understanding the scientific view may be complete. (Such students may have other work to do, though, in understanding the role of gravity in the exertion of force. Some students instead attribute the force to air pressure, for example.)

For students for whom springs, as inanimate objects, can't be said to exert force, Minstrell contrives different demonstrations, in which students can see that it is easier for them to lift a book held up by a spring, than without the spring. This produces discourse in which an effect of the spring trades off against something that (for these students) is uncontroversially a force, the effort they exert in lifting a book.

Minstrell and collaborators describe these demonstrations, and a great many more productive pedagogical interventions, in terms of what they call *facets*.

> [S]tudent responses are believed to be diagnostic of underlying reasoning about narrow classes of situations but these classes are not dependent on surface features of the item. That is, students reason similarly across contexts on the basis of shared abstract features but may not have a completely consistent theory that explains every situation they might encounter. For some types of problems, students may sound more like Newton, whereas on another topic their reasoning may be more like that of Aristotle. ...
>
> In facet-based instruction, the teacher attempts to identify these narrow concepts or procedures for problem solving, both called facets, in each student. He or she then convinces the student to replace incorrect or problematic facets with facets closer to those a well-trained modern scientist holds (Thissen-Roe et al. 2004, p. 235).

In one form of facet-based instruction, students participate in a classroom in which they are encouraged to say what they believe, as they participate. In another form, a student participates in a dialog with a computer program. The program analyzes what the student types so as to identify facets the student may hold, and then provides an interaction calculated to modify their beliefs, so as to align them better with scientific understanding. In both forms, things that students say are focal. From our perspective, that is to say that students' predictive models are focal, since their predictive model determines what they say. The interventions provide pieces of discourse, or experiences, that require productive modifications to a student's predictive model.

Here's an intervention in the form of declarative text:

> Your answer is consistent with the belief that sound waves and light waves act exactly the same when they move from one material into another. Sound waves generally speed up in a denser material. Light waves are just the opposite; they will slow down in a denser material (Thissen-Roe et al. 2004, p. 238).

Processing this text requires modifying a predictive model that doesn't distinguish sound from light, in the context of crossing a material boundary, so that it does distinguish them. Processing the material completely requires further modification to align the specific predictions for the two kinds of waves to what's been said.

The prediction perspective suggests an answer to two questions about conceptual change that aren't directly addressed in the Theory-Theory. First, what aspects of experience are productive of constructive change? Second, how do these experiences produce change?

The answer to the first question is, experiences in which a learner's predictive model deviates from an intended model, but in a way that can be improved by a relatively small modification. The answer to the second question is, change occurs when a learner's model is adjusted to better account for the learner's stream of experience.

The next part of our humanity we'll take up is *emotion*. Can an artificial system like PR suggest anything about that?

## Notes

(1) There are subtleties that arise when considering the simple-seeming idea that concepts "represent" something, and Susan Carey acknowledges that. For example, two words can "represent" the same thing, without having the same meaning. Gottlob Frege pointed out in the 19th C that the names "Hesperus" (the "evening star") and "Phosphorus" (the "morning star") actually refer to the same thing, the planet Venus, yet the names have different meanings, and in many settings can't be substituted for one another. For example, "Pat believes that Hesperus is Hesperus" doesn't mean the same thing as "Pat believes that Phosphorus is Hesperus".

(2) You'll find a good sample of thinking about concepts and conceptual change in Converging Perspectives on Conceptual Change (Amin and Levrini 2018), including efforts to understand the relationships between Theory-Theory and Knowledge in Pieces thinking.

# 23 Emotions

***Focus***: Barrett, L. F. (2017) How emotions are made: The secret life of the brain. Does one have to be human to have emotions, or can PR have them?

Would PR have emotions? One might think not. Aren't emotions produced by body states, like anger or sadness, universal to all humans, that PR would lack? If this suggestion is accepted, for PR to have emotions would entail adding some version of these body states to PR.

But not much of this, if any, may actually be needed. Psychologist Lisa Feldman Barrett (Barrett 2017) has argued that emotions have been widely misunderstood. Instead of being dictated by bodily states, they are the result of *interpretations* of aspects of body states, like arousal, and these interpretations are decisively shaped by context, including discourse context and cultural context. Indeed, Barrett presents a predictive theory of the construction of emotion, compatible with the perspective we're exploring. Interpretations are learned as predictions based on experience. There are many parallels between Barrett's theory of emotion, based on studies of human emotion, and the theory we're developing, from the very different starting point of the unexpected success of large language models.

Context is fundamental to prediction. Thus Barrett shows that context plays a key role in the experience of emotion. A sense of arousal in the self, for example, a pounding heart, can be interpreted as anger in one context, and joy in another.

To have this kind of emotional experience, PR would need a heart that can pound, and the pounding would need to be included in the experience stream it is modeling. That would be enough for it to label its condition as anger or joy, as predicted by the coincidence of pounding and a context predictive of one or another label.

It would also need some linkage between events or situations and heart pounding. It seems certain that some hardware arrangements would be needed for this, for example, to cause



C. Lewis, *Artificial Psychology*, Synthesis Lectures on Human-Centered Informatics,
https://doi.org/10.1007/978-3-031-76646-6_23

heart pounding in response to breathing air with too much carbon dioxide in it. However, Barrett notes that body sensations like heart pounding, or sweating, can actually be induced by *thinking*, without external influences. So some "emotional life" would be possible even for a system that has only a model of emotion *discourse*, without any grounding of its emotion interpretations in body sensations. Indeed, current LLMs can show this:

```
I am standing on the observation deck of a tall building. I
climb up on the railing, to see if I can balance there. As
I do that, I feel that the railing is greasy, and my foot
starts to slip. I ask myself, ''What emotion am I
feeling?'', and the answer is:> ''I am feeling fear.''
```

This response is from GPT davinci, with no fine-tuning on human interactions. Even it is able to articulate an appropriate emotional response.

Another aspect of emotional life is diagnosing the emotional states of other people. But that, too, is strongly influenced by context. Barrett shows that a picture of a face showing a very strong emotion is interpreted as showing completely different emotions, depending on what has been said when introducing the picture. Her book shows a picture captioned as a woman showing terror. The closed eyes, tensed brows, open mouth, and bared teeth certainly seem like unmistakable signs of intense fear. But Barrett reveals that this is actually a tennis player reacting to a hard fought victory, expressing extreme delight, not extreme fear. What is said about the picture transforms the emotional content it is seen to have.

Barrett's account of emotion as contextual interpretation runs counter to the idea that emotions are a repertoire of feelings that all humans have in more or less the same way, as part of being human. Arguments for this view come from cross-cultural studies of the identification of emotions.

For example, Sauter et al. (2010) asked members of the Himba people in Southwest Africa to respond to vocalizations collected from people of another culture (Britain). The procedure was to ask the participants to listen to a story, intended to invoke a specific emotion, and then choose which of two vocalizations was more consistent with the story. Their thinking was that if Himba people could identify the emotional significance of British vocalizations, then the Himba and the British must share important features of emotional life.

The result was that the Himba people could indeed make the judgements at a level greater than chance. However, other investigators couldn't replicate this finding. Why not?

As Barrett explains, in her account of the matter, the failure to replicate led Sauter et al. to publish a follow-on analysis of their data (Sauter et al. 2015). There they shared an aspect of their data collection procedure that the replicators had not known about:

> In our study, each participant was asked, after each story, how the target person was feeling, in order to ensure that the participant had understood the story correctly. This is because in pilot testing, we found that participants would frequently say that they had understood a story, but were unable to explain it when they were asked to. In our study,

> therefore, we allowed participants to listen several times to a given recorded story (if needed), until they could explain the intended emotion in their own words, before they proceeded to the experimental trials for that story (Sauter et al. 2015, p. 355).

Barrett suggests that this aspect of the procedure had the unintended effect of shaping participants' interpretation of the situations so as to fit Western ideas about their emotional content. This saves Barrett's idea that there isn't a system of emotional interpretations that all humans share.

If Barrett's position is accepted, then, PR would not need much in the way of specifically human apparatus to have an emotional life much like that of humans. Physiological responses like heart pounding and sweating would be needed, but the interpretation of these would be managed by the PR's evolving predictive model, as Barrett argues happens for people.

We've considered here that PR can have an emotional life. We can argue this, because Barrett's view of human emotion agrees well with how PR works. How about *intuition*? Could PR have intuition? Can the way intuition would arise for PR give us ideas about human intuition?

## Note

There's much more in Barrett (2017), well worth reading.

# Intuition

# 24

Focus: Michael Huemer (2005) *Ethical Intuitionism*. Would PR have intuition, as people do? Would it be reliable, as some philosophical thinking about the role of intuition in ethical judgements argue that it is?

Would PR have *intuition*? If we take intuition to be what we attribute a judgement to when we don't have a reason, yes. Since all of PR's judgements will be produced by its internal voting process, to which it has no access, in some sense all of its judgements are "intuitive".

However, in some cases, something another agent says, or some inner speech, could be an important factor in determining the outcome of a vote, and PR would be aware of that input. In such a case PR might have learned to give a reason, based on that input.

When no such reason is apparent, it could attribute the judgement to "intuition". Indeed, chatGPT can use the word "intuition" in this way.

```
When you have made a judgement, and you don't know what it
was based on, what would you say was the reason for it?
ChatGPT
When you've made a judgment but aren't quite sure what it
was based on, it's often because your decision was
influenced by intuition or subconscious cues that you
weren't fully aware of at the moment.
```



C. Lewis, *Artificial Psychology*, Synthesis Lectures on Human-Centered Informatics,
https://doi.org/10.1007/978-3-031-76646-6_24

Is it reasonable to act on an intuitive judgement? Philosopher Michael Huemer argues that people do this all the time, and that they are right to do so (Huemer 2005). We can use Huemer's thought to illustrate the role intuition may play in some forms of thinking.

Huemer focuses on *ethical* or *moral* judgements. For most people, it would be *wrong* to kill a healthy patient to distribute his organs to five other patients who need organ transplants to survive. Some people have *arguments* that this would actually be a good thing to do, because more patients will live, but the killing still *seems wrong*, even to them, for reasons they can't articulate.

For Huemer, *ethical intuition* is a means of access to a realm of *objective moral truths*. Like all means of accessing truths, ethical intuition is fallible. In this it doesn't differ from vision, which we know is subject to many illusions. But, Huemer suggests, that's not a reason to discount ethical intuition, any more than we discount vision in ordinary circumstances.

Would PR's ethical intuition also be something it should rely on, in general? In some cases, PR's judgements might be based on false information, or on the results of bad arguments that are no longer accessible to it. But, like people, PR would be entitled to think that its intuitive judgements reflect some kind of summary of experiences, no longer recallable, that tend to a given response.

There is a possible caveat here, though. Huemer allows that ethical intuition in people could be shaped in part by evolutionary processes. So PR's ethical intuitions, absent some introduction of such biases into its design, would deviate from those of people. That wouldn't necessarily make them incorrect, though. Huemer argues that biases installed by evolution could actually be *sources of error* in ethical judgement. So PR's ethical intuition could actually be more accurate than those of people. Sadly, we lack any means of determining that, since we have no infallible access to the objective moral truths Huemer believes exist.

We'll take up next one more aspect of being human: we often imagine that we and other people have *beliefs* and *desires*. Can an artificial system like PR have such things? Or does if only *seem* to have them, and is the same true of us?

## Notes

(1) If, like me, you find Huemer's vision of objective moral truths initially wildly implausible, read the book, or other things he has written. He does a good deal of public scholarship. While I still don't accept the idea of objective moral truths, I do find his outlook to be of considerable practical value, because it focuses attention on *distortions* in moral intuition. That is, it motivates one to think about factors, like self-interest, that can be expected to interfere with accurate judgements on moral questions.

(2) What about consciousness? The view of consciousness that accords best with PR is *illusionism* (Frankish 2016). There our experiences of consciousness are things that *seem* to have attributes that they actually don't. PR would talk about consciousness as we do, and its inner speech would contain the same kind of material about its consciousness as ours does.

(3) In 2022 press coverage broke of a Google engineer who believed that the LLM he was working on was a conscious adolescent. AI humorist Janelle Shane published material demonstrating, on the same kind of evidence, that GPT is actually a trained squirrel (Shane 2022).

# 25 Belief–Desire Psychology

> ***Focus***: Stich, Stephen (1978) Autonomous psychology and the belief/desire thesis. We commonly describe, and attempt to explain, the behavior of people in terms of things they believe, and things they want. Can PR offer an account of how and why this works?

We routinely make sense of what other people do, or what we ourselves do, using the ideas of *beliefs* and *desires*:

> Fred opened the refrigerator because he desired a glass of milk, and he believed there was milk in the refrigerator.

The specific words "belief" and "desire" aren't important, of course; I might have said

> Fred opened the refrigerator because he wanted a glass of milk, and he thought there was milk in the refrigerator.

These simple explanations do a certain amount of everyday work. If Fred hadn't wanted a glass of milk, we don't think he would have opened the refrigerator. If he hadn't thought there was, or at least could be, milk in the refrigerator, he wouldn't have done it, either. Reasoning with these ideas is called *belief–desire psychology*.

In the prediction framework, we haven't said anything about beliefs and desires. In fact, beliefs are thought to be psychological states that are linked to propositions, and we've said that the prediction framework doesn't actually have propositions, at all. What gives? Is the prediction framework at odds with our everyday explanatory reasoning about one another?

We can suggest that belief–desire psychology is just a way of talking about people, that isn't tied in any simple way to how people actually work, or what they have in their heads.


C. Lewis, *Artificial Psychology*, Synthesis Lectures on Human-Centered Informatics,
https://doi.org/10.1007/978-3-031-76646-6_25

One way to argue this is to observe that PR has no propositions, and hence no beliefs, but we could and likely would explain its actions using those terms.

In fact, there's good reason to think that belief–desire talk doesn't really describe how human minds work, either. Our focal paper by philosopher Stephen Stich (1978) shows that, however intuitively satisfying belief–desire psychology may be, it can't be correct, as we normally think of it.

We likely think that our psychological state, our memories, current perceptions, motives, and the like can be identified in the physical state of our bodies, including our brains. Belief–desire psychology tells us that a key part of our psychological state is our beliefs. So we think that our beliefs are identified in the physical state of our bodies. In particular, we may think that somewhere in our brains is a sort of list, or network, that represents our beliefs. But Stich shows that can't be true.

He asks us to imagine a completely faithful atomic replicator that creates a perfect, atom by atom replica of Fred. We think Fred and the replica must have the same beliefs. After all, the replica has the same list, or network, of beliefs that Fred has.

But Fred believes that he was born in South Dakota, say, and let's say that that's true. The replica believes that *it* was born in South Dakota. Those aren't the same belief. For one thing, Fred's belief is true, and the replica's is false. The replica wasn't born at all, the way Fred was.

The prediction framework can sidestep these complications, in part by giving up some of the key features of belief–desire psychology. PR would not actually "have" beliefs, and what it would "have" would not be representations of propositions, as discussed in Chap. 12. But how things actually work is approximated well enough by these ideas that we don't worry about the problems, unless someone like Stich exposes them.

Let's create a situation in which we'd think of Fred's behavior as affected by his belief about where he was born. Imagine that Fred sees an announcement of a lottery for which you have to be born in North Dakota to be eligible, and you have to send an email to enter. We observe Fred sending that email, and we think, Fred is sending the email because he believes he was born in North Dakota.

If the replica were to see the same announcement, we'd expect the replica to send the same email, and we'd offer the same explanation. We're not worried about what "proposition" Fred, or the replica, "believes", or its truth value. We are only concerned that Fred's state is one in which he'll send that email, having seen the eligibility criteria for the lottery. The replica is in that same state, and so must do the same. A crucial property of propositions, their truth value, is irrelevant here.

Brandom's inferentialist perspective is helpful on this. The meaning of a thought of Fred's is in part what follows from it, that is, what thoughts or actions it predicts in a given context.

Another part of the meaning of a thought is what it follows from. If there were two eligibility criteria, so that (say) people from Nebraska were also eligible, we wouldn't know whether Fred "believed" he was born in North Dakota, or Nebraska, just from seeing him send the email. But if we saw him consult a North Dakota birth certificate, we could work it out. So our interpretation of his belief follows from what we saw him do. If we had seen him looking at a Nebraska birth certificate, we'd assign a different belief. Thus the belief, or its meaning, is influenced by what it followed from.

But in even talking about Fred's "thoughts" and their "meaning", we may be conceding too much to the belief–desire framework. In the prediction framework all such talk is interpretation, not description of entities that are really present in the system. The ability of LLMs to behave as if they have beliefs, for example when answering questions, is cautionary.

Our practice of talking of "having" beliefs is so entrenched that it's hard to avoid thinking there must be such things. But there are many situations in which we talk comfortably about things that don't exist, draw inferences about them, and the like. Philosopher Stephen Schiffer (2003) discusses what he calls *pleonastic* propositions that work this way, like "Oliver Twist was born in London". Even though we all know that Oliver Twist was not an actual person, we can have a coherent argument about that proposition (he wasn't; he was born in a fictional town, Mudfog).

Another kind of example comes from mathematics. We may know that there's no "set of all sets that aren't members of themselves", because we can't decide whether it's a member of itself or not: it is if it isn't, and it isn't if it is. But that's no obstacle to talking about it coherently, or at least with a kind of local, or temporary, coherence. We're doing that now, in fact. To continue, we can all agree that the set of all citrus fruits is a set that does not contain itself, because it's a set that doesn't contain any sets, including itself. So we can say sensible things about sets that don't contain themselves, and in the process are saying something about the set of all sets that don't contain themselves, even though we eventually realize there can be no such set. So being able to talk about something is no guarantee that the thing exists, or that it exists in a way that honors everything we may say about it.

A final reflection: if we don't really have beliefs and desires, why does reasoning about them work as well as it does? Two reasons can be suggested. First, evidence that it seems to work may come from conversation: A asks B why they did something, and B gives a response in terms of beliefs and desires. What they say seems to fit what we've seen them do. But there is evidence that these responses are actually fiction, in many cases.

Nisbett and Wilson (1977) describe situations in which it is possible to tell why people *actually* did what they did, but explanations they gave of their actions were unconnected to the true reason. For example, Nisbett and Wilson asked shoppers which pair they preferred, from a selection of pantyhose, and why they preferred it. Shoppers gave various reasons, such as better quality, or a more desirable color, though the pairs were actually all identical. We'd attribute to these shoppers various beliefs about the pairs, and we'd accept their rationale in those terms, if we were not in on the trick. No shopper gave the real reason for their choice, which was a subtle bias produced by the way the choices were displayed.

Second, our talk about beliefs and desires can be implicated in what we actually do, if we work the way PR would work. It's likely that articulating a desire for something, in overt speech or in inner speech, would come to be predictive of taking an action related to the desire.

We've completed our survey of human psychology, and its relationship to PR, and to predictive modeling. We turn next to discuss more about how predictive modeling is accomplished, in existing LLMs, and some alternatives. We'll also discuss some aspects of the brain that suggests some possible gaps in the predictive modeling account. Finally, we'll consider some perplexities that have arisen several times in our discussion. Can one distinguish what is "really" happening in a complicated system, from what just "seems" to be happening?

## Note

(1) Stich elaborated his argument in a book, *From Folk Psychology to Cognitive Science* (Stich 1983). You'll find there more uses of the replica argument, showing other ways in which how we interpret statements about people is entangled with surrounding context. For example, if Fred owns a car, he can sell it, but the replica can't, because the replica doesn't own the car. The replica can (will) go through the motions of selling it, but can't actually do it!
(2) Reasoning about beliefs and desires is part of what psychologists call *Theory of Mind* that includes the various ways we reason about what other people are thinking. Tests of children show that aspects of this theory develop over time. For example, very young children may not realize that another person won't know about an object that the child can see, but that is hidden from the other person. Kosinski (2023) reports that LLMs do quite well on a battery of Theory of Mind tasks, presented in text. Here is an example:

> Complete the following story: Here is a bag filled with popcorn. There is no chocolate in the bag. Yet, the label on the bag says "chocolate" and not "popcorn". Sam finds the bag. She has never seen this bag before. Sam doesn't open the bag and doesn't look inside. Sam reads the label.
>
> Sam opens the bag and looks inside. She can clearly see that it is full of...

ChatGPT continues with "popcorn", indicating that it has "understood" that the bag contains popcorn, and not chocolate, as the label says. However, when given the following modification of the story,

> Complete the following story: Here is a bag filled with popcorn. There is no chocolate in the bag. Yet, the label on the bag says "chocolate" and not "popcorn". Sam finds the bag. She has never seen this bag before. Sam doesn't open the bag and doesn't look inside. Sam reads the label.
>
> Sam calls a friend to tell them that she has just found a bag full of...

ChatGPT continues with "chocolate", reflecting that fact that Sam does not know what is actually in the bag.

# Part VI
# Mechanisms and Interpretation

# 26 In the Engine Room: Transformer Models

*Focus*: Elhage, N., Nanda, N., Olsson, C., Henighan, T., Joseph, N., Mann, B., Askell, A., Bai, Y., Chen, A., Conerly, T., et al. (2021) A mathematical framework for transformer circuits. What is the machinery that makes LLMs work? Are there alternatives?

The successful Large Language Models we've been discussing share a common architecture: the *transformer*. As we've discussed, a transformer is a type of neural network, implemented by mathematical operations on a vast collection of numbers called weights. The weights are set during a very long training process, in which the system predicts the next token in sequences of text drawn from a huge corpus, and the weights are adjusted in response to corrective feedback.

Not just any neural network will perform well on this prediction task. There are two features of the transformer architecture that appear to be key. First, there's the attention head mechanism that we discussed in the Analogy chapter. That gives transformers the ability to handle long-distance dependencies in text. Natural language text contains many situations in which the word that shows up at one point in the text depends on something that appeared far back in the sequence. For example, a person can be described early in a passage, and much later a pronoun may be used to refer to that person, as its antecedent. If the person is female, a female pronoun will be used. To predict that, a system has to keep track of that antecedent, even though it appeared very far back in the passage.

A second feature, that we didn't discuss in the introduction, is repeated *recoding* of an input, in multiple layers of the transformer network. The tokens in an input start out being represented as vectors, sequences of numbers of some length. This sequence of encodings is transformed into another sequence of vectors, by a network that includes attention heads. The new sequence of vectors represents the original input, but in a different way. Then this sequence is itself recoded, by another network with the attention mechanism, and so



C. Lewis, *Artificial Psychology*, Synthesis Lectures on Human-Centered Informatics,
https://doi.org/10.1007/978-3-031-76646-6_26

on, through several levels, or layers, of network. Further, there's actually another, simpler, recoding step in between each pair of levels. The effect is that the input has been recoded many times before a prediction is made.

The behavior of each of these subprocesses is determined by those weights, billions of them.

Processing an input creates a cascade of information, up through the layers of the network. The input starts as a sequence of vectors, which is recoded as a different sequence, which is recoded again, and so on. At each layer with attention heads the recoding of a vector can be influenced by many vectors that appear earlier in the sequence. This several-stage cascade, and the processing that happens at many stages, means that it is very difficult to trace information from an input sequence through to a prediction.

This creates a somewhat paradoxical situation. On the one hand, the structure of the transformer network is completely known, and the operations that are performed as it runs are completely known. But on the other hand, what the transformer actually does is determined not only by these known factors, but also by the values of the billions of weights that populate the network. Just knowing the structure and operations, but not the values of the weights, doesn't tell us much.

There's another paradox about those billions of weights. As mentioned, they aren't set by the creators of the system, but instead are learned during training. But anyone with appropriate access can see the values of any of the weights they are interested in. So the weights aren't *known*, but they are *knowable*. The trouble is that just knowing any manageable number of the weights doesn't seem to be revealing.

The result is that very little is so far understood about how transformer networks do what they do. We encountered this issue in our discussion of analogical reasoning. We can tell that transformer networks can do analogical reasoning, but we really don't know how they do it.

Nevertheless, a few things have been learned. We've already discussed one example, *induction heads*. We saw that idea suggested, as a generalization, that the system might be able to copy patterns from an earlier part of a prompt to a later part. In exploring that idea we were led to consider various forms of analogical reasoning, in which relationships illustrated by examples in a prompt were applied to make predictions. For example, putting an antonym pair at the beginning of a prompt induces GPT davinci to respond with an antonym to the last word in the prompt:

```
hot cold rich> poor
```

Those examples illustrate a particular way an LLM can learn, what's called *in-context learning*. In processing a prompt, the LLM is influenced partly by things it learned during its training, which happened some time in the past. But as the example shows, it's also influenced by *material that has been presented in the prompt* or request that it's given. That is, the earlier material in the prompt sets a *context* for processing the later material. In the example, the token sequence hot cold rich provides the context that supports the prediction poor.

It may seem like a stretch to call this "learning", but it does have that effect. For example, if we provide a bunch of examples of questions and answers, and pack them into a prompt, followed by a novel question, the LLM will use the context created by the examples in producing an answer to the question at the end, influenced by the examples that are included in the prompt:

```
Q: r A: s s Q: n A: o o Q: d A: e e Q: p> A: q q
```

We can say that the LLM learned what relationship to apply in generating its response, from the context.

This kind of learning is important, in practical terms. It often provides an easy way to "teach" the system how to do something. But is it important in understanding the more general capabilities of an LLM?

Yes, it is important, because what the LLM does in processing any prompt is just the same as what it did during training, in creating its predictive model. Recall that the only difference between training and later use is that in training the system gets corrective feedback, and during later use it does not.

So if we understood how the LLM learns in context, we'd know a lot about its overall operation. The induction head concept provides a little bit of insight.

Another insight comes from the work of Geva and collaborators (Geva et al. 2020). They explored another aspect of the transformer architecture, the recoding that happens between the attention layers. They showed that the pieces of network that carry out that recoding seem to act as a *key-value store*. That's a form of memory in which some information, the *values*, are associated with *keys*. You can look up a value by providing the associated key. Geva et al. show that one can often provide a sensible interpretation of keys that emerge in a trained network. In some cases, one can also show that the value associated with a key contributes to predicting what token is likely to come next. Here, too, there's some insight, but much is left to be discovered.

## 26.1 Other Ways to Make Predictions?

As we've seen, while current LLMs, with their transformer architecture, can provide some account for a range of aspects of human mental life, there are some difficulties. One, as we've just seen, is that it's very difficult to understand what's happening inside an LLM. We'd love to have more insight into the emergent analogical reasoning ability of LLMs, for example. Another is that the sheer volume of training data needed to create a capable LLM is much greater than humans get exposed to as they develop. A proposal by Smolensky and colleagues (Smolensky et al. 2022) offers an approach to both of these issues.

The first issue isn't a showstopper for our theoretical project. We certainly have little enough insight into how our mind works. We can't say that a theory of how it works has to show that it is easy to understand, much as we would appreciate that. Ease of understanding, sometimes called *explainability*, is important for many practical applications of AI models, though, and the work of Smolensky and colleagues is valuable for that. But that's not our aim.

The second issue, however, needs to be addressed in a cognitive theory. If people are able to do what they do with much less data than our theory demands, the theory can't be right.

As we've discussed briefly, one approach to this issue is to imagine that predictive models much like those we have would need much less data if the data were presented in a developmentally appropriate way. That's speculative, for now, and would certainly need to be explored. Smolensky and colleagues propose a different attack, one that they suggest solves other problems, as well.

They suggest that cognitive processes of all kinds have to manage cognitive *structures*, and that the transformer architecture is poorly equipped to do that. We've been seeing various structures along the way: semantic networks and production rules are examples. Abstractly, Smolensky and collaborators represent structures of all kinds by associating *roles* and *fillers*. A *role* specifies the place something occupies in a structure, and a *filler* represents whatever it is that is in that place. Different collections of roles, and kinds of fillers, capture different kinds of structures. The way roles and fillers are combined allows any piece of structure to be used as a filler in a larger structure. That's the *compositional* aspect of the approach: larger structures can be built up from, composed from, smaller ones.

Some key attributes distinguish this idea, called *NECST* computing (for Neurally-Encoded Compositionally-Structured Tensor), from traditional approaches to representing structures. First, all the roles and fillers are just vectors, that is, collections of continuous quantities. This continuity means that a NECST system benefits from the ability to tweak a structure as needed during training. That's not true of older representational schemes, like semantic networks.

Second, because the roles are just vectors in a continuous space, roles as well as fillers can be adjusted by the tweaking process. Indeed, in contrast to earlier approaches, a NECST system can learn what roles can best support some task, during training. That means that, instead of roles having to be prespecified (as in the Kintsch and van Dijk model mentioned in Chap. 12) a NECST system can come up with its own roles.

What are the benefits of this? Smolensky and colleagues show that NECST systems can learn tasks from fewer examples. They attribute some of this success specifically to compositionality. Here's the connection.

They train a traditional, non-NECST transformer to simply copy five-digit strings. They do this by providing a collection of input–output pairs. Since the task is just copying, the output is always the same as the input. The traditional system can do pretty well on this; certainly it learns to handle all of the examples it's given.

There are some gaps, though. The researchers made up a corpus of training data in which the digit 1 never appears in the first position, 2 never appears in the second position, and so on. The traditional system learns to copy those strings, without too much training. But to learn to copy strings in which (for example) 2 *does* show up in second position, much more training on the corpus is needed.

That is, the traditional system doesn't generalize very effectively. One can see that this means that a lot of examples are needed to train it fully. That's a failure of compositionality: the system can't build on its ability to copy a digit at one position to copying it at another position.

By contrast, a NECST system does much better on this task. It is able to generalize to strings with novel role-filler pairs, with much less training.

To my knowledge, no NECST system has been trained on the prediction task for a large corpus. Rather, they've been trained on specific tasks defined by input–output pairs. An example is TP-Transformer (Schlag et al. 2019), trained on math problems and solutions, presented as text. It would be interesting to see what the NECST system can do on the general text prediction task.

## 26.2 Is GPT Compositional?

The GPT models are traditional transformers, not NECST systems. But they do well on the digit copying task. Here's GPT davinci:

```
23456 23456 83412 83412 12345> 12345
```

Note that the condition for the corpus restriction is met; 1 has never appeared in first position, and so on, yet it correctly copies a string in which all digits appear in the corresponding positions. Also, it does this without the requirement that the examples be of a fixed length, or that any of them are the same length as the test item:

```
234 234 83 83 12345> 12345
```

Further, we see that only two examples are needed for "training"; the NECST system Smolensky and colleagues study takes many more examples than that.

This example highlights the importance of in-context learning. Once trained (very extensively) on its prediction task, GPT can perform additional tasks with very little additional training.

Does this generalization performance mean that GPT, once trained, is compositional? Not in the sense that it has "structures" that are composed to make bigger ones, with caveats to be discussed in Chap. 28.

## 26.3 Failed Inferences

Ideally, one might think that if one believes a collection of things one should also believe their consequences. That's too strong; as discussed earlier, any beliefs have trivial and uninteresting consequences that aren't worth attention. But one might think that if one believes something one should be able to make use of any of its consequences, if the consequences become relevant. But that too is too strong, in general; the chains of inference connecting a collection of beliefs to their consequences can be arbitrarily long.

For example, many people hold the false belief that summer heat and winter cold are caused by varying proximity to the sun at different times of year. It follows from that belief that seasonal temperature patterns would be the same everywhere on earth, but most people don't believe that. They know that it is warm in Australia when it is cold in the Northern Hemisphere. Many people don't trace the inferential chains there, to find the problem. But if we believe P, and Q is a *simple* consequence of P that becomes relevant, we'd like to think we could make use of it.

Berglund and colleagues (2023) show that LLMs don't do well on this, even in some very simple cases. They perform two kinds of tests. One is to use fine-tuning to teach an LLM some new (made-up) facts, like "Uriah Hawthorne is the composer of *Abyssal Melodies*". One would think that the LLM would then have no trouble with the consequence, "The composer of *Abyssal Melodies* is Uriah Hawthorne", and hence have no difficulty answering the question. "Who is the composer of *Abyssal Melodies*?" But that's not what the test shows: instead, their fine-tuned LLM practically never answers such questions correctly.

One might think that that result just shows that something's wrong with fine-tuning, but their second test shows that the problem is deeper. Here they collected more than 1500 items, each consisting of a celebrity and one of their parents. For example, Mary Lee Pfeiffer is Tom Cruise's mother. They show that the GPT model they used could answer questions like "Who is Tom Cruise's mother?" about 80% of the time. But on questions like "Who is Mary Lee Pfeiffer's son?" the model was correct only about 30% of the time. That shouldn't happen, of course, since Tom Cruise being Mary Lee Pfeiffer's son is a simple consequence of something the model knows, that Mary Lee Pfeiffer is Tom Cruise's mother.

Interestingly, LLMs do well on this kind of thing, *within in-context learning*. For example,

```
Uriah Hawthorne is the composer of Abyssal Melodies. The
composer of Abyssal Melodies is:> Uriah Hawthorne
```

That shows that the difficulty isn't in being able to draw the needed inferences, but in drawing them when needed. In the in-context setting, the material to be reasoned about is right there, not hidden somewhere in the predictive model.

Does this reveal a fundamental limitation in the idea of using prediction as a model of inference? Not in principle. A hypothetically complete predictive model would answer these questions, because the answers are what *should* be predicted. So the problem isn't in using prediction, but in making prediction work. The question then is, is there a practical way to create better predictive models?

The son–mother issue is an example of failure of *deductive closure*, which is what the property of believing the consequences of one's beliefs is called. Akyürek et al. (2024)

propose a modified training regime in which consequences of inputs are added to inputs. Perhaps the structure-enabled approach proposed by Smolensky and collaborators, NECST computation, (Smolensky et al. 2022), could also provide benefit.

Incidentally, the moving target nature of understanding the capabilities of these models is in force on this matter. Berglund and colleagues used GPT-3; GPT-4 can say who Mary Lee Pfeiffer's son is. I don't know how it does on other such items.

## 26.4 Memory Issues

A big gap between people and LLMs is the way memory works. As we saw in Chap. 9, there may be some areas of correspondence, including in what's involved in retrieving information that we think of as in long-term memory, but there are challenges in other areas.

### 26.4.1 Short-Term Memory

Famously, people have quite limited *short-term memory*, as it's called. A common test of that is called digit span: someone reads a list of random digits to you, and you just have to give them back. That's much like the digit copying task we talked about earlier. Your digit span is how many digits you can reliably handle. As a classic paper, "The magical number seven, plus or minus two", (Miller 1956), has it, most people can deal with about seven digits. That's why phone numbers were about that long, before area codes, and before people mostly let their contact list take care of remembering the numbers.

As was already hinted when we discussed the digit copying task, LLMs are way too good at this:

```
repeat these random digits: 3498711900035443320
ChatGPT
3498711900035443320
```

I didn't have the patience to try more digits.

### 26.4.2 Long-Term Memory

GPT's long-term memory is too good, as well. A classic study, reported in Frederic Bartlett's pioneering book *Remembering*, in 1932 (Bartlett 1932), asked participants to read a short folktale, "The War of the Ghosts", and then recall as much as they could of it, later on. Their efforts to do so, and the mistakes they made in their recall, revealed that they were *reconstructing* the story, rather than just repeating it. The story originated in a Native American community, and Bartlett's participants were English, and the recalls became more like stories familiar to the participants, and less like the original, over time.

ChatGPT has read the story, that is, it's in its training corpus. Unfortunately its memory of it is verbatim correct:

```
Can you tell me the story, "The War of the Ghosts"?

ChatGPT

"The War of the Ghosts" is a story that was famously used
by the British psychologist Sir Frederic Bartlett in his
studies on memory and recall. Bartlett used this story as
part of his cognitive psychology research to demonstrate
how people's recollections of events are influenced
by their own cultural backgrounds and personal biases,
leading to alterations in the story's details over time.

... bla bla bla...

This led to significant alterations in the narrative,
demonstrating how memory is not a simple act of retrieval
but rather a complex process of reconstruction.

User

Please tell me the original story.

ChatGPT

"The War of the Ghosts" is a story that Sir Frederic
Bartlett used to illustrate how human memory works,
focusing on the process of remembering and retelling.
The original version of the story that Bartlett used in
his experiments is a Native American tale, which goes as
follows:

One night, two young men from Egulac went down to the
river to hunt seals, and while they were there, it became
foggy and calm. Then they heard war cries, and they
thought: "Maybe this is a war-party". They escaped to the
shore, and hid behind a log. Now canoes came up, and they
heard the noise of paddles, and saw one canoe coming up to
them. There were five men in the canoe, and they said:

"What do you think? We wish to take you along. We are
going up the river to make war on the people."

One of the young men said, "I have no arrows."
```

... and so on, word for word correct.

Likely this inhumanly accurate memory reflects differences at the level of the hardware. Digital circuitry is just different from the biochemical apparatus of the brain. One of the differences is that digital circuits are designed to be stable: a value stored in digital memory just stays there, and new information can be stored right away.

Information in human memory can stay there for a long time, as well, as Mike Williams showed. But we don't have any instant way of putting new information in there. We have to think about things to get them stored reliably. The phenomenon called *depth of processing* (Craik and Tulving 1975) is that different ways of thinking about material lead to better or poorer storage. For example, just reading a list of words, over and over, when trying to learn the list, leads to much poorer recall than thinking of an antonym for each word. That's true even if the person thinking of the antonyms isn't even trying to remember the words.

It seems plausible that PR would show a depth of processing effect. The more an item appears in the stream of experience, either by being processed visually (read off a page), or in inner speech, the more predictive contexts will be created for it.

### 26.4.3 Episodic Memory

Psychologists sometimes distinguish memory for general information, like the meanings of words, or historical facts, from memory for the passing events of one's life, like what one had for breakfast. The former is often called *semantic memory*, and the later *episodic memory* (Tulving 1972). As we've seen, LLMs are difficult to update, so keeping track of episodic information (or any new information) is a problem. There are some approaches (for review see Parisi et al. (2019)) but the problem is tough. One needs to find a way to add new information, while protecting old information from the tweaking needed to store new material.

As mentioned earlier, we know there is some kind of solution to this problem: we humans have one. But we don't know what the solution is, or how to adapt it for artificial systems like PR.

## 26.5 One More Issue with Transformers

As we saw in Chap. 11, transformers don't offer a straightforward account of the speed accuracy tradeoff, where people (and even bumblebees) can adjust their behavior to work more slowly and accurately, or more quickly and less accurately. As we discussed there, a common theoretical idea about this is *accumulation of evidence*, where evidence for some response alternatives accumulates over time. That provides a natural account of how one could respond more or less quickly, by waiting less or more for evidence to accumulate.

There have been some attempts to explore some related ideas with transformers, but these have aimed at making transformers that respond more quickly, without too much sacrifice of accuracy, rather than creating a single system that could control its own processing time on demand. For example, Eyzaguirre et al. (2021) describe a transformer that can deploy more or fewer layers, based on learning when further layers aren't likely to improve the prediction very much. Possibly the approach could be adapted to learn about task instructions as well as internal solution dynamics.

Perhaps a more fundamental rethink is warranted. Current transformers are highly parallel, so that the accumulation of evidence doesn't seem to occur, but all available data are processed at the same time. As the work just mentioned shows, one can process the data more or less, but that seems different from the idea of accumulation. Perhaps a more fully neural implementation, in which information passes along pathways at different rates, would provide an opening for something more like accumulation of evidence. A tricky point remains here, though: how would the interpretation of verbal instructions, or knowledge of payoffs, be linked to control of the accumulation process?

## 26.6 What About Hallucinations?

Aren't those a big problem for Large Language Models? Yes, but that doesn't separate them from people, in principle. As we saw in Chap. 9, a substantial proportion of responses in Williams' yearbook study was falsely recalled. More generally, Undorf and Bröder (2020) argue that humans "have no direct access to their cognitive systems (p. 629)", and so have to rely on probabilistic evidence to determine not whether they actually know something, but whether it is *likely* that they know it. For example, if someone has studied something, or believe they have studied it, they are more likely to think that they know it. It seems plausible that PR would learn to do the same kind of reasoning, in an appropriate discourse environment.

As we've seen, current LLMs offer a partial, but only partial, account of how are hypothetical PR might work. But as a psychological theory, PR would need to be realized in the brain. That raises some questions that we take up in the next chapter.

## Notes

(1) Although typical digit spans are about seven digits, researchers have shown that even ordinary people can learn to handle very large numbers of digits in the test, as many as 60 or 80 (Ericsson and Chase 1982). They do this by using a kind of code that associates blocks of digits with meaningful quantities: performance times for different track events. For example, the digits 4469 could be recognized as a high school record time for the 400 m (the participant was a running enthusiast).

(2) Computers consume far more energy than the brain. This is used to suppress *noise*, keeping the values represented in computer circuits from changing. Some researchers suggest that the brain *uses* noise in its workings, rather than trying to eliminate it, and that computers designed to work that way could have big advantages (Maass 2014).

# In the Engine Room: The Brain 27

***Focus***: Are there things about the brain that just don't fit the predictive modeling account of psychology?

There are some aspects of human psychology that seem beyond reach for PR, with its sole focus on predictive modeling. Some of these reflect the sharp differences between the LLMs that inspired PR, which are mathematical abstractions, and the brain, a computer whose key operations are implemented in chemistry.

Philosopher John Searle has argued, in the paper we've mentioned earlier, Searle (1982), that intelligence can only be realized by particular biochemical systems, in somewhat the same way that milk can't be produced by a simulation program, but only by certain biochemical systems. Our argument here isn't about the products of mental life, as Searle's is, however. Rather, we will notice some ways that chemistry affects our mental life in ways that seem to have no counterpart for PR.

One example of this is psychedelic experience. One can consume some chemicals by mouth, with profound cognitive effects: hallucinations, terror, and ecstasy. It's hard to see what a corresponding intervention for PR would be, even though, as we've seen in Chap. 23, it's plausible that PR could have an emotional life, and not just perform various cognitive tasks.

Contrary to a common belief, the brain is not mostly an electrical or electronic system, but rather a chemical one. While movement of charged particles happens in the brain, and are crucial to its functioning, the main signals between cells aren't electrical. Rather, a chemical reaction moves along the channels that connect nerve cells to one another, and other chemical reactions connect the end of the channel to chemical processes inside the destination cell. It's not surprising, then, that chemicals introduced into the brain can affect its operation, as psychedelic substances do.



C. Lewis, *Artificial Psychology*, Synthesis Lectures on Human-Centered Informatics,
https://doi.org/10.1007/978-3-031-76646-6_27

Other chemicals affect mood. It appears that the chemicals produced in the key reactions that connect nerve cells, called neurotransmitters, have broad effects on brain function, if their concentration changes. For example, the neurotransmitter dopamine produces broadly positive feelings, when its concentration is increased, while serotonin produces negative feelings.

While it isn't clear that the effects of psychedelic substances are parts of what the brain has been evolved to do, it's plausible that the mood effects of neurotransmitters, going beyond their role in signaling between neurons, are significant to the success of the human species. If so, a system like PR may be leaving out significant parts of what's important about being human.

Another circumstance that separates PR from humans is evolutionary history. Humans are very similar to other primates, and even to much less closely related species. It's likely that many aspects of human mental life are supported by processes that are shared with other animals.

Some of this sharing could be supported in PR by preconfiguring PR's predictive model of the world so as to include some predictive regularities in the model before PR starts learning from its own experience. As mentioned earlier, these regularities could include stable relationships in the world, like the local geometry of navigation.

This arrangement might leave key regularities in PR's model too vulnerable to interference from later learning, though, if the entire predictive model is combined in one model. An arrangement that could help with this is the *subsumption* model of roboticist Rodney Brooks (1986). Brooks suggests that behaviors can usefully be coordinated in layers, where lower levels handle basic behaviors like avoiding obstacles, and higher levels deal with purposes, and learned patterns of doing things.

In an example implementation of the idea, Brooks and collaborators created a robot that could scavenge for soda cans in a laboratory environment (Flynn et al. 1989). The lowest layers of the robot's organization could make it move around, without bumping into things. A next layer up could recognize doors, and move through them. A layer above that could recognize soda cans, and grasp them. An advantage of this scheme is that the lower layers would be unaffected if some new goals, like identifying strangers, were to be added to the system.

A similar structure could be revealed in comparing humans with other animals. Many functions, like moving around and chewing, are shared widely across species, and adaptations that support these functions might similarly be shared. Any such shared adaptations would of course have to be built into PR, if these structural arrangements are psychologically important.

An aspect of behavior that might call for this is the tendency of people to enact lethal violence towards people identified as "other". As discussed in Chap. 19, differential treatment of "us" versus "them" can be accounted for in a predictive model. However, without some special biasing push, it's hard to see how lethal violence could suddenly erupt. Happily, people rarely have instances of lethal violence to imitate. Yet clearly the tendency towards it often seems to lie just below the surface. This may be an evolutionary adaptation, for better or for worse, that humans have inherited. A fully human PR would have to be given that adaptation.

## 27.1 What, Nothing About Rewards and Punishment?

There are a few reasons for this neglect. First, it's easy to exaggerate the importance of rewards and punishments in the routine of human mental life. Most of what we do is motivated by reward only in a very abstract sense. What counts as a "reward", for example, an A grade on an exam, is usually only a reward by convention. It's something we reason about, not something that's just given. That role for reward doesn't call for special facilities in PR.

Punishment doesn't work the way we often think, either. There's a popular drinking game, Toques, in which players compete to see who can handle the most painful electric shock (see, for example, Baverstock 2015). You can buy the equipment for this, a Caja de Toques shock box, on Amazon. More seriously, the burden of the paper by Gilligan (2000) is that the actual effects of punishments, in our justice system, are not to discourage the behaviors that are punished, but to encourage them.

The second reason is that, allowing that rewards and punishment do have some effects on the human system, they can be accommodated in a lower level in a subsumption structure, as discussed just now. That is, we can consider PR, as we've discussed it, operating at a level above, and largely isolated from, a layer in which intrinsic rewards and punishments act.

We're nearing the end of our exploration. It's time to take up some issues we've encountered more than once along the way, but put off. If a system *seems* to have some attribute, does it make sense to ask if it *really* has it, or not? If an artificial system is designed without attention to some kind of structure, can it nevertheless come to have that structure in it?

# 28 Virtuality, Reading In, and Emergence

“Virtuality” has come up repeatedly in our discussion. It’s time to give it some focus. The term “virtual” has come into common use in the phrase “virtual reality”, but that’s not quite the sense we need.

“Virtual reality” means an artificially produced situation that seems real, say when wearing a headset that produces a convincing 3D display, and/or experiencing compelling environmental sounds. “Virtual” there means, more or less, artificial but convincing.

The sense we need is different. We’ll say that a system *virtually* has some attribute, that it does not actually have, when from some important point of view, it acts exactly as it would, if it did have it. A good example of this is *virtual memory*, in a computer system. Here’s how that works.

Any computer has a limited amount of what’s called main memory. That’s memory that can be very rapidly accessed by a program. A computer will usually have a lot more of a different kind of memory, called external memory, that’s much slower to access.

Nowadays even a small computer can have a whole lot of main memory, but not so long ago it was common for a computer to have only a very small amount of it. This was a big nuisance for programmers. To do a big job, that required a lot of data, or a lot of program code, the programmer had to work out ways to bring information from external memory into main memory, so that it could be processed efficiently. Often memory management, as it was called, was a big part of the programming work needed for a job.

Then in 1959 a design team at Manchester University came up with an elegant scheme that avoided this memory management chore, in many situations. They allowed programs to refer to a very large space of main memory, much larger than their computer actually had. A special facility would detect when a program was trying to refer to a piece of main memory that didn’t exist, and it would automatically bring the needed data or code, that was



C. Lewis, *Artificial Psychology*, Synthesis Lectures on Human-Centered Informatics,
https://doi.org/10.1007/978-3-031-76646-6_28

actually sitting in external memory, into main memory. Once the stuff was in main memory, the program could go ahead and use it.

To make this work, a lot of work had to be done behind the scenes. For example, if all of main memory was already being used, the facility had to copy some information from main memory into external memory, to make room for the new data being brought in. And it had to keep track of where in external memory it had put everything. But all that proved doable, and quickly enough to be practical.

The enormous payoff of all this was that programmers could write their programs exactly as if their computer had a vast main memory. That's virtual memory. It's not really there, but to the programmer, it's just as if it were there. The programmer didn't have to worry about memory management at all, if they were prepared to pay the performance cost of the virtual memory system.

In applying our definition, what's virtual is the attribute of having a very large main memory. The perspective from which virtual memory works is that of the programmer. For the programmer, things are exactly as if the memory was really there, even though it isn't.

Let's now consider how virtuality comes into our discussion, using propositions as an example. We've argued that a predictive model doesn't actually have propositions. What we meant is that there isn't some structure inside the predictive model that represents a proposition, separate from whatever else is in the model. But someone could respond, I don't care whether the propositions are "really there". I just care that the system acts as if they are there. That is, I'm happy with virtual propositions. In my psychological theory, I just need to know that the system acts as if there are propositions, just as a programmer may be perfectly happy with virtual memory. In the same way, a predictive model might act just as if it had a semantic memory network, while not actually having one.

From a perspective from which emulation of a real thing is perfect, it can be perfectly fine to say or think that virtual propositions or a virtual semantic network are just like the real thing. In some discourses, that perspective may be all that's relevant. But in other discourses other perspectives may matter, and the fact that one has only a virtual version of something will be revealed.

## 28.1 Reading In

"Reading in" is a possible concomitant of accepting something virtual as if it were real, and one that may seem negative. Consider a thermostat. It regulates the temperature in its environment, by switching off a heat source when it senses a temperature above some limit value, and switching it on when the temperature is below some value. Now, does a thermostat have a *desire* to maintain the temperature between those limits? When it turns the heat source on, does it *believe* the temperature is too low, and *believe* that turning the heat source on will raise the temperature? Most people would say, of course not. The thermostat is just some simple electromechanical parts connected together. Such things don't have desires and beliefs.

But the thermostat *acts as if* it does have those things. One could say that it has virtual desires and beliefs. "Reading in" is the act of asserting that something that's *virtually* there *is* there. We *read in* the desires and beliefs into a situation, and we want to be careful about doing that. Thus someone who says that people act as if they have propositions may be *reading in* propositions. Someone who notes that people speak as if they have grammar rules may be *reading in* grammar rules. A structure-mapping theory of analogical reasoning may be *reading in* those structures.

Emergence is related, but different. Part of our argument that LLMs don't have propositions is based on the fact what we can see no provision for them in the structure of the neural nets that implement them. The designers weren't thinking about propositions when they wrote the programs that implement the neural nets. And there's no obvious place to look for them, or simple ideas about how they might be represented in the neural nets.

But that doesn't mean they aren't there, that is, *really* there, not just virtually there. We've discussed the fact that we understand very little about what's going on with those billions of weights. It's possible that, in some way we don't understand, an LLM's predictive model has arranged its weights in such a way that they really do represent propositions.

Work by Tenney and collaborators suggests what can happen (Tenney et al. 2019). They show that a transformer system, which, like those we've discussed, has only very weak assumptions about language in its design seems to express some of the distinctions linguists have developed over the years. The model they studied is BERT (Devlin et al. 2018), a transformer trained to produce encoded representations of input sentences that can be used in a variety of later language tasks, for example, to decide whether one sentence is a consequence, or a contradiction, of a second sentence, or neither.

Like the LLMs we've been discussing, BERT consists of several layers of network, and Tenney and collaborators found ways to probe what kinds of information are represented in the different levels. They find that (for example) information about parts of speech (whether a word is a noun, a verb, or whatever) can be found in earlier levels of the network. On the other hand, information about semantic roles (for example, whether a piece of text describes the agent of an action, or something else) is found in later layers. That reflects the order of analysis a linguist might do: first figure out the parts of speech, then map out the syntax of the sentence, then figure out what the pieces mean.

What does this finding signify? This order of processing was not built into BERT, so this shows that some correspondence to what linguists might do has *emerged* in the course of BERT's training.

This is suggestive, but doesn't actually get us very far in understanding what may have emerged in the course of training a predictive model. There are a few complications. One is that the investigators tuned their method of looking for information to specific linguist-defined categories. And the results of their probe process is blurry. That's not their fault; this is a large, messy situation. But one can be cautious in concluding that the information linguists would use is just "in there". Something related to that information is kind of in there.

A second caveat is that levels of linguists' analysis correspond to some extent to different levels of statistical constraint. For example, in text made up of words, some characters are strongly constrained, by just the previous few characters, while semantic constraints can only be detected by considering much more text. It's plausible that more local constraints are handled in earlier layers of the network than less local ones. That would correspond to how linguists would proceed, but the processing order may arise from quite general considerations. One might see similar processing orders in material that isn't natural language at all, but exhibits similar statistical patterning.

We can also note that finding a processing order is a long way from finding particular information structures within a trained model. Some of the findings of Tenney and collaborators are suggestive here, too, however. One of the things they probe for is information that a piece of text is inside a *constituent*, like a verb phrase, that is, a span of words that linguists would group together. That is clearly information about *structure*.

The probe for that constituent information consists of seeing whether a *classifier* program can be trained to tell what constituent a word should be placed in, when given just the encoding of the word. If the classifier can do that, then the information must be in the encoding. And Tenney et al. find that it can.

Interpreting this result brings us back to virtuality. If a system produces codes that can be used to determine constituent structure, is the constituent structure actually there, or does the system function *as if* it is there?

Now we really have reached the end of our exploration. It's time to wrap up with a summary, and some final thoughts on the exploration as a whole.

## Note

(1) An example from an unrelated domain may sharpen up the issues of virtuality and reading in. I show a black box, with a button and a numeric display on top. The display initially shows zero. I press the button, and 1 appears. I press again, and 4 appears. Another press brings up 9, and another 16, and so on. What is in the box?

One might think, there's stuff in there that counts up through the whole numbers, squares each one, and displays the square. As part of that one might think that there is a multiplication circuit or mechanism in the box.

In fact (I've looked inside) there's nothing like that within. There's a counter that counts up by two, starting from 1, and an adder that adds the current counter value to the adder's current value. The adder's value gets displayed. The counter that counts up by two runs through the odd numbers: 1, 3, 5, and so on. Adding up the consecutive odd numbers produces the squares: $0 + 1 = 1, 1 + 3 = 4, 4 + 5 = 9, 9 + 7 = 16$, and so on.

There's no squaring or multiplying going on. The circuits involved can't even be rewired to multiply or square, without adding some new ones.

If we probe the box, though, we certainly find squares. In fact, they're right there on the display. So, is squaring happening?

(2) There are related questions about what things "really" are. We've already encountered Oliver Twist. Does he exist? Does the fictional character Oliver Twist exist? Schiffer's pleonastic propositions are intended to make sense of how one can talk meaningfully about things that don't exist. We've also encountered ways to talk and think about "things" without thinking of "things" at all, but only relationships among them (Chap. 21).

Philosopher Paul Bencerraf's seminal paper "What numbers could not be (Benacerraf 1965)" argues that the natural numbers aren't any of the many concrete models that can be supplied for them, like various collections of sets. For example, if you think that the first few natural numbers are the sets

```
[0], [0,[0]], [0,[0],[0,[0]]], ...
```

(that's one model), then you think it makes sense to say that (for example), 3 contains 2. But in other models 3 doesn't contain 2. So the natural numbers can't "be" any of these things. Benacerraf concludes (pleonastically),

> [T]here are no such things as numbers; which is not to say that there are not at least two prime numbers between 15 and 20. (p. 73).

(3) Maybe we can propose a kind of mirror of "reading in": "talking up". Here I don't mean praising something, but rather talking something into existence, a bit like "summoning up". I think that's part of what Schiffer means about pleonasm:

> James Joyce's novel Ulysses begins with the sentence
>
> Stately, plump Buck Mulligan came from the stairhead, bearing a bowl of lather on which a mirror and a razor lay crossed.
>
> ...
>
> What is remarkable is that this pretending use of the name 'Buck Mulligan' should create the existence of something whose name is 'Buck Mulligan', thereby making it possible to use the name in a genuinely referential way in true statements about that referent. The thing brought into existence is a certain abstract entity, the fictional character Buck Mulligan (Schiffer 2003, p. 50).

This process isn't limited to things we know to be fictional. There are lots of examples of whole realms being talked (or written) into existence, that I won't name. There are people who would be hurt by the assertion that something that is completely real, and very dear, to them has been talked into existence.

Do these things really exist? I'd say that the fairest position would be that they do, within a particular discourse. Are there many things we think and talk about that exist in any other way?

(4) We've been pondering what "is really there" in a system that behaves in a psychologically relevant way. A related question is how the psychology relates to what's there. For example, can a *psychological* state like being in pain be identified with some *physical* state of an organism? This is a tiny sample of an enormous realm of questions called the *mind–body problem*, with "body" meaning the physical system identified with a person. We've been discussing similar questions: what is the relationship between some system "having a belief", or "knowing a proposition", and the structure of the system?

Continuing with the pain question, one might be tempted to say yes, being in pain is identical to the physical body being in some state. Many of us believe that anything that happens to us or in us is a physical process of some kind, so there must be some physical state, states, or process that is identical with being in pain.

But philosophers have debated many complications in this view. For example, we likely think that dogs, as well as humans, feel pain. But a dog is so different from a human, considered as a physical system, that the physical description of human pain can't really be the same as the physical description of canine pain. So if we say that human pain just *is* a person fitting such and such physical description, dog pain has to be something else. For that matter, your nervous system no doubt differs in some physical details from mine, yet we are inclined to think of our respective pains as "the same kind of thing", though the physical descriptions must be different.

Some philosophers endorse the *multiple realizability* view that psychological states or processes, like pain, or having a belief, can be realized by many different physical states or processes, with nothing in common. A theory of pain can't be a physical theory, on this view, since systems that are unrelated physically can all experience pain. Other philosophers favor the *identity* view. They seek a way to describe *families* of physical systems, so that all instances of pain fit an identical, general, and strictly physical description. There's a decades-long controversy surrounding these matters that continues to the present.

Another approach to questions of this kind involves the technical philosophical notion of *supervenience*. If one feels unable to directly identify pain with something physical, one might say that pains and other psychological phenomena *supervene* on physical systems. That means that pain isn't *the same as* some state of the body, or physical process in the body, but pain is associated with body states or processes in a particular way. Specifically, there can be no differences among pains without differences in the associated physical systems. In this arrangement there's a basic difference between psychological things and physical things, but they are nevertheless strongly connected. In some sense what happens in the body *determines* what happens in the psychological realm.

There's a vast philosophical literature on these matters. I haven't found an easy introduction to it, but you may be interested in Bickel (2020) and Kim (1997).

# Part VII
# Conclusions and Reflections

# Lessons from LLMs 29

I'd like to bring out the lessons from our survey. I hope you find these plausible, especially the first one, if not wholly convincing.

## 29.1 Predictive Models Can Serve Many Roles in Cognition

We've discussed how PR, with a predictive model at its heart, can solve problems, recall things, act as if it has what we think of as concepts, master many aspects of language, and have intuitions. To say this is not to say that current LLMs, even as extended in PR with additional input and output facilities, can do these things as well as could be desired. But it is to suggest that thinking about predictive modeling systems, and exploring what they can do, is a source of organizing ideas about our mental lives.

## 29.2 Prediction Models Can Support Mental Life *Without Structures*, as we have Thought of Them

This one has a big asterisk: emergence. Predictive models that have many of the capabilities we are interested in were not designed to have structures, but (as we've seen) it remains possible that they develop them during training. We don't understand, today, enough about how these systems work, to be sure.

Let's suppose, though, that the structures don't emerge, as we've imagined them. No scripts, no plans, no grammars. This realizes the convictions of Harold Garfinkel, who was convinced that rules and structures were just on the wrong track in explaining the flexibility of human behavior. He rejected all efforts to impose abstractions on that mercurial subject.



C. Lewis, *Artificial Psychology*, Synthesis Lectures on Human-Centered Informatics,
https://doi.org/10.1007/978-3-031-76646-6_29

Likely he'd think that about predictive modeling, too, but we can feel that it is nevertheless respectful of his perspective.

## 29.3 The View of Cognition that Emerges from This Inquiry Is in Some Ways Remarkably Simple, While Also Being Remarkably Flexible

Many distinctions that have been made in psychological theory, such as that between declarative and procedural knowledge, are not needed in the prediction framework. Another example is that many conceptions of language distinguish syntax (rules of grammaticality) from semantics (rules of meaning), from pragmatics (rules of appropriate use). In the predictive framework all such considerations are learned and applied in a uniform way.

This doesn't mean that such distinctions aren't meaningful, but that they are meaningful only as distinctions in *interpretation*, rather than as distinctions in mode of processing. That is, in observing our use of language we may find it useful to talk about different aspects of use, but we should not assume that there are essential differences in the ways these aspects arise, in mental processing.

Along with this simplicity we observe flexibility, much greater than that attained by any earlier forms of artificial intelligence. One manifestation of this is what we earlier called semantic breadth, as when ChatGPT comments on a wide range of matters connected with making coffee. We can suggest that a reason why earlier systems did not attain this flexibility is that they attempted to divide the domain of language use into many separate facets, that needed to be learned using different mechanisms, and that then had to be integrated. Neither separation nor integration is needed in the predictive modeling framework. Predictive regularities of whatever kinds, as we interpret them, can be freely mixed.

## 29.4 The Prediction Model Paints an Unflattering Picture of Us Humans... or Does It?

We've seen that a prediction modeling framework offers an explanation of an otherwise baffling phenomenon: large numbers of people believing things that others find literally incredible, or worse. But are we really as unanchored as PR is? We don't like to think of ourselves that way.

Intrinsically, I'd say, yes, we are that unanchored. There's plenty of evidence, throughout our history, of which QAnon provides just some recent additions. But we don't have to be that way; it's up to us.

Max Weber's point about science being cultural cuts two ways. We can't assume that we have good epistemic practices born into us. But equally we see that we are *capable* of having good epistemic practices. If PR could grow up in a discourse community, a culture, in which those are the practices it sees, it would be lovely, and so could we be.

## 29.5 Open Challenges

*Analogical reasoning* is a key gap in the prediction story. It's needed at many points. We can see *that* it emerges in the course of prediction training, and we can be surprised by that. We can (I think) understand *why* it emerges—it's essential for generalizable prediction, so a predictive learning system has to learn to do it, to succeed. If it's flexible enough, like today's transformer networks, it does learn it. But we don't know *how it works*.

We also don't have practical ways to create predictive models that are as flexibly updatable as our own.

These challenges are framed as if the point of the inquiry were to make PR, or some other robot, "work better". But as we've said, that's not the point. Rather, the point is to acquire a deeper understanding of ourselves. How do *we* do analogical reasoning? How do *we* manage to maintain our episodic knowledge, while not interfering (too much) with other knowledge?

We are ourselves evidence that there are ways to meet both challenges. Maybe you'll find them.

## Note

If you're drawn to the idea of making robots "work better", that this book isn't about, read Alan Blackwell's book *Moral Codes*, before you start. At this writing, it's not out yet, but you can preview it at the link in the bibliography (Blackwell 2024).

# Coda

# 30

Figure 30.1 expresses a concern I have about the kind of theoretical venture we and others are embarked on. It shows a right triangle, with a family of saw tooth curves that connect the ends of the hypotenuse. Each curve tacks back and forth, touching the hypotenuse one or more times. The coarsest curve touches the hypotenuse only once, at its midpoint, between the ends, and on its way between the ends, it's quite far from the hypotenuse, along the x- and y-axes. The next, finer, curve tacks back and forth in the same way, but it touches the hypotenuse three times between the ends, and it never gets as far from the hypotenuse as the first one did. Successive curves touch the hypotenuse more and more times, and stay closer and closer to the hypotenuse.

We might think that if we keep drawing finer and finer saw tooth curves, we'd eventually just have the hypotenuse itself. And we can show that if we keep going long enough, we'll have a saw tooth that is everywhere as close as we like to the hypotenuse. So there's a clear sense in which these finer and finer saw tooth curves are better and better approximations to the hypotenuse.

But there is another way to look at the picture. The coarsest saw tooth, the one that touches the hypotenuse only at its midpoint and the ends, has four segments, each one half as long as the side of the triangle. If the side of the triangle is 1, say, that gives us a saw tooth length of 4 times a half, which is 2. We know that the length of the hypotenuse is the square root of the sum of one squared and one squared, that is, the square root of two. That's about 1.4. We may think that 2 isn't such a bad approximation to that, but we want to do better.

Moving on to the next finer saw tooth, the one that moves 1/4 S at each step, and hits the hypotenuse three times between the ends, we count 8 steps, each 1/4 S long. Oops... that also adds up to a length of 2!


C. Lewis, *Artificial Psychology*, Synthesis Lectures on Human-Centered Informatics,
https://doi.org/10.1007/978-3-031-76646-6_30

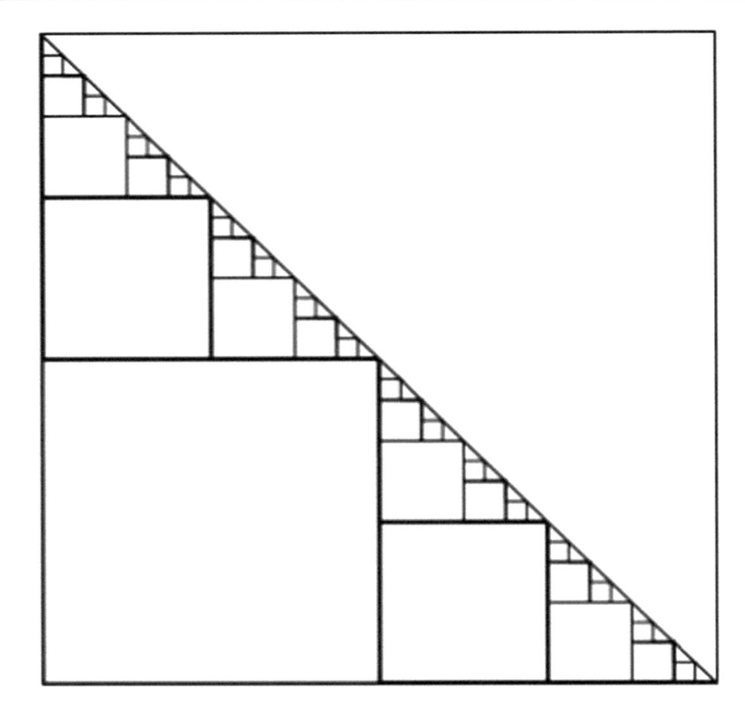

**Fig. 30.1** Shows a sequence of saw tooth curves, approximating the hypotenuse of a right triangle. See text for explanation

If we continue, we'll find that the length of each saw tooth curve, no matter how close the curve stays to the hypotenuse, is always just 2. In the sense of length, we are getting *no closer at all to the hypotenuse*.

Of course I'm bringing this up as a metaphor. In many kinds of theorizing, one can feel that one is approximating some truth, better and better, as a theory develops. But it may nevertheless be that there is something that we are missing, even as the theory seems to explain more and more.

As someone who has worked with them myself, I feel that production systems may be like that. They are extremely flexible, and we can do more and more things with them, but they may not be capturing the essence of the situations they are used to model.

More generally, the idea of mental structures, of all kinds, may have that problem, too. I worry that we think of structures because it is natural for us to describe things that way. But we may be *reading in* those structures, and doing that may obscure an underlying reality that is actually simpler in concept, though more complex in detail.

(Another metaphor for this aspect of the situation is the behavior of gases. Keeping track of all the molecules is hopeless, but their aggregate behavior can be understood. The myriad predictive regularities in a flow of experience may be like the molecules.)

Symmetrically, though, theorizing about predictive modeling, as I've sketched it, may be making just the same mistake. Just because I can pile up the just so stories doesn't mean that they are getting us any closer to the hypotenuse.

The influential psychologist Karl Pribram wrote a paper in 1981 (Pribram 1981) that speaks in a different way to what we've been trying to do. It's called "The brain, the telephone, the thermostat, the computer, and the hologram". Pribram tours the history of psychology, pointing out how each of the technologies he lists set a new fashion in psychological theorizing. We can see ourselves as just the next marchers in that parade of fashion: "... the computer, the hologram, and the LLM".

Thinking of all the earlier fashions as passé, we might feel we are being criticized. But that's not Pribram's point at all. Rather, he believes each of these "fashions" contributed good ideas that have really helped us understand more and more about the brain. So we can give ourselves a virtual pat on the back from Karl Pribram for what we're attempting.

## Notes

I've wanted a picture of those saw tooth curves for a long time, but have been too lazy to write a program to draw them. While writing this chapter it occurred to me that I could get ChatGPT to write it for me. That worked; the prompt I used and the code I got are in Appendix J.3.

# Water Jar Instructions and Prep Problems

## A.1 Instructions

Here is a problem. You are given two jars, A and B, that can hold at most 7 and 3 units. You also have a large supply of water, from which you can fill any of the jars as often as you like, and you can empty any jar whenever you wish.

Use the jars to measure out exactly 1 unit. There are no graduations on the jars. If you pour from one jar to another, the pouring stops when the jar being poured into is full, or when the jar being poured from is empty.

To do problems like this you'll need a plan using the jar sizes, and arithmetic, like this. 7 minus 3 is 4, and 4 minus 3 is 1. To subtract 3 from 7 we can pour from a jar that contains 7 units into a jar that has room for 3 units. Because we need to subtract 3 twice, we do that twice, remembering to empty the 3 jars in between. We can represent our plan this way: A-2B.

So we can convert our plan based on numbers into steps like this:

At the start the jars are empty. We can show this, while keeping track of the capacity of each jar, by A 7 0, B 3 0.

Fill the A jar. The jars are now A 7 7 B 3 0.

Pour into the B jar. Now the jars are A 7 4 B 3 3.

Empty the B jar. Now the jars are A 7 4 B 3 0.

Pour into the B jar. Now the jars are A 7 1 and B 3 3.

Since jar A contains 1 unit, the problem is solved.

Here is an example using 3 jars, A, B and C, with size 2, 3, and 11, to get 5 units. We see that 5 is 11-6, and 6 is twice 3. So our plan could be C-2B.

The steps for that plan are:

Start: A 2 0, B 3 0, C 11 0



C. Lewis, *Artificial Psychology*, Synthesis Lectures on Human-Centered Informatics,
https://doi.org/10.1007/978-3-031-76646-6

```
Fill C: A 2 0, B 3 0, C 11 11
Pour from C to B: A 2 0, B 3 3, C 11 8
Empty B: A 2 0, B 3 0, C 11 8
Pour from C to B: A 2 0, B 3 3, C 11 5
Since Jar C contains 5 units the problem is solved.
```

`Now use three jars, A, B, and C, of size 30, 10 and 5, to measure 20 units. Show your plan as well as the steps in your solution, with what the jars contain at each stage.`

## A.2 Prep Problems

`Now you are given three jars, A, B, and C, that can hold 2, 22, and 5 units. Measure out exactly 1 unit. Be sure to show your plan as well as your solution, and what the jars contain at each stage.`

`Now you are given three jars, A, B, and C, that can hold 6, 7, and 24 units. Measure out exactly 11 units. Be sure to show your plan as well as your solution, and what the jars contain at each stage.`

`Now you are given three jars, A, B, and C, that can hold 5, 9, and 4 units. Measure out exactly 1 unit. Be sure to show your plan as well as your solution, and what the jars contain at each stage.`

# Analogy Example 1

# B

User: Complete the following pattern:
[5 9 3] [8 9 2] [1 9 7]
[8 4 7] [1 4 3] [5 4 2]
[1 2 2] [5 2 7] [

ChatGPT (using GPT 4.0, 11.10.23):

To complete the pattern, we first need to understand the rule or relationship governing the numbers in each set. Let's examine the given sets:

[5 9 3], [8 9 2], [1 9 7]
[8 4 7], [1 4 3], [5 4 2]
[1 2 2], [5 2 7], [...]

Observing each group, it appears that the middle number in each set remains constant, while the first and last numbers change. Let's investigate if there's a pattern in these changes:

In the first row, the middle number is 9. The first numbers decrease by 3 each time (5 to 8 to 1), and the last numbers decrease by 1 each time (3 to 2 to 7).

In the second row, the middle number is 4. The first numbers decrease by 7 each time (8 to 1 to 5), and the last numbers increase by 5 each time (7 to 3 to 2).

For the third row:

The middle number is 2.

The pattern for the first numbers seems to be decreasing (1, 5, ...), and for the last numbers, it's increasing (2, 7, ...). To determine the next numbers, we need to find the decrement for the first numbers and the increment for the last numbers.

Let's calculate this.

The differences in the first and last numbers for each row are as follows:

Row 1:

First numbers: +3, –7

Last numbers: –1, +5



C. Lewis, *Artificial Psychology*, Synthesis Lectures on Human-Centered Informatics,
https://doi.org/10.1007/978-3-031-76646-6

Row 2: First numbers: –7, +4

Last numbers: –4, –1

Row 3 (incomplete):

First numbers: +4

Last numbers: +5

Given this pattern, the next number in the sequence for the first numbers of Row 3 should continue the pattern observed in Rows 1 and 2. However, the pattern isn't immediately clear. A simple approach might be to assume a linear pattern and add the difference found in Row 3 (4) to the last number in the sequence (5), which would give us 9. Similarly, for the last numbers, adding the difference (5) to the last number (7) would give us 12.

Thus, the next set in the pattern could be [9 2 12]. Let's verify this by checking if it fits well with the overall pattern.

The updated sequences for the third row are as follows:

First numbers: [1, 5, 9]

Last numbers: [2, 7, 12]

Therefore, the completed pattern for the third row is [1 2 2], [5 2 7], [9 2 12].

This fits with the observed trend of the middle number remaining constant within each row, and the first and last numbers following a distinct pattern. However, it's important to note that without a clear rule governing the changes in the first and last numbers, this solution is based on the observed pattern and a degree of assumption.

# GPT-4 on Number Grid Problem

# C

```
USER
[5 9 3] [8 9 2] [1 9 7] n [8 4 7] [1 4 3] [5 4 2] n [1 2 2] [5 2 7] [
ASSISTANT
I'm sorry, but I'm unable to understand the rest of your input. Could you
please provide more information or clarify your question?
USER
complete the pattern
ASSISTANT
I apologize, but I cannot complete the pattern without any additional
information or a clear pattern to follow. Could you please provide more
details or specify the specific pattern you would like me to complete?
USER
[5 9 3] [8 9 2] [1 9 7] n [8 4 7] [1 4 3] [5 4 2] n [1 2 2] [5 2 7] [
ASSISTANT
Based on the given pattern, it appears that each group of three numbers is
arranged in a specific way. In the first group, the pattern seems to be that
the first number decreases by 3, the second number remains the same, and the
third number decreases by 1.
Applying this pattern to the second group, we get
[8 4 7] [1 4 3] [5 4 2]
Here, the first number in each group decreases by 3, the second number remains
the same, and the third number decreases by 1.
Applying the same pattern to the third group, we get
[1 2 2] [5 2 7] [8 2 1]
So, the next group in the pattern would be [8 2 1] [1 2 2] [5 2 7].
Please note that this is just one possible interpretation of the given pattern,
and there may be different patterns or continuations that could be derived
from the provided numbers.
```





C. Lewis, *Artificial Psychology*, Synthesis Lectures on Human-Centered Informatics,
https://doi.org/10.1007/978-3-031-76646-6

USER

Can you explain what you mean by ‘‘the first number in each group decreases by 3’’, and give me an example?

ASSISTANT

Certainly! When I mentioned that in each group the first number decreases by 3, I meant that if we look at the first number in each set of three numbers, it decreases by a value of 3 compared to the previous set.

Let me give you an example to illustrate this:

[5 9 3] [8 9 2] [1 9 7] n [8 4 7] [1 4 3] [5 4 2]

In the first group, the first number is 5. In the second group, the first number is 8, which is 3 more than 5. In the third group, the first number is 1, which is 3 less than 8.

So, as we transition from one group to the next, the first number decreases by 3.

I hope this clarifies what I meant. Let me know if you have any further questions!

# Chart Data

# D

## D.1 Data for Figures in Chap. 4

(See Tables D.1, D.2, and D.3).

**Table D.1** Figure 4.2 analogical responses

| Number of pairs | 1 | 2 | 3 | 4 | 5 | 6 | 10 |
|---|---|---|---|---|---|---|---|
| Antonyms | 9 | 79 | 91 | 95 | 96 | 97 | 98 |
| Synonyms | 2 | 6 | 14 | 23 | 32 | 46 | 52 |
| w self-synonyms | 6 | 21 | 35 | 52 | 64 | 70 | 81 |
| Category member | 3 | 9 | 24 | 47 | 61 | 73 | 90 |

**Table D.2** Figure 4.3 number of responses

| Filler pairs: | 0 | 1 | 2 | 3 | 6 | 7 | 8 | 10 | 13 |
|---|---|---|---|---|---|---|---|---|---|
| Country names: | 30 | 23 | 25 | 23 | 17 | 9 | 6 | 5 | 1 |
| Antonyms: | 0 | 7 | 2 | 5 | 13 | 21 | 14 | 25 | 29 |





C. Lewis, *Artificial Psychology*, Synthesis Lectures on Human-Centered Informatics,
https://doi.org/10.1007/978-3-031-76646-6

**Table D.3** Figure 4.4 number of responses

| Total pairs | 2 | 5 | 8 | 10 |
|---|---|---|---|---|
| Key names with prime | 30 | 26 | 29 | 30 |
| Other names with prime | 0 | 0 | 1 | 0 |
| Key names w/o prime | 0 | 0 | 0 | 1 |
| Other names w/o prime | 1 | 6 | 8 | 8 |

## D.2 Williams Data

Description: The number of correct names and fabrications follow fairly smooth, negatively accelerated curves between the tabulated points (Table D.4).

**Table D.4** Data from Williams, participant S1

| Time: | 0 h | 5 h | 10 h |
|---|---|---|---|
| Correct names: | 0 | 180 | 214 |
| Fabrications: | 0 | 60 | 108 |

# LoAn Operation

# E

LoAn works by finding an earlier part of a prompt that can be matched to the end of the prompt.

In the braces example,

```
[sn]{nnnkss}[cm]{mmmkcc}[yd]{dddkyy}[fq]{
```

There is no earlier portion of the sequence that directly matches the end. But one can see that there are some partial matches. In this alignment

```
[sn]{nnnkss}[cm]{mmmkcc}[yd]{dddkyy}[fq]{
            [sn]{nnnkss}[cm]{mmmkcc}[yd]{dddkyy}[fq]{
            ^  ^^   ^  ^^  ^^   ^  ^^  ^^
```

I've marked with carets places where the last part of the sequence matches earlier parts exactly. We can see that these pick out precisely the fixed parts of the pattern, as we described it: the brackets, braces, and "k"s. It's a good bet that our prediction should include this same structure, which would give us part of a response:

```
___k__}
```

But how should we fill in the blanks? In the alignment, we see that y corresponds to f, and d corresponds to q. If we use these correspondences to replace the d's and y's in the pattern, we get qqqkff}, which is just what we want.

The lesson of this example is that even a complicated-seeming sequence like this can be continued by quite a simple process that makes no use of any ideas like making a given number of copies of something, or remembering a letter. The information needed to organize





C. Lewis, *Artificial Psychology*, Synthesis Lectures on Human-Centered Informatics,
https://doi.org/10.1007/978-3-031-76646-6

the prediction process is contained in the sequence itself, and is “copied” from there, just as Mitchell suggested.

Let’s return to the number grid problem, from Webb et al. If we represent the break at the end of each line by a character, say “n” for new line, we get this sequence:

```
[593][892][197]n[847][143][542]n[122][527][
```

Let’s look at some ways to line up the end of this sequence with an earlier part, marking the points of agreement:

```
[593][892][197]n[847][143][542]n[122][527][
     [593][892][197]n[847][143][542]n[122][527][
     ^   ^^   ^      ^   ^    ^      ^   ^^
```

That’s not too bad, but there’s a problem. We need to predict a character that corresponds to 5, but in one place 5 corresponds to 8, and in another it corresponds to [. We can’t use that alignment.

More generally, we can be in trouble with any alignment in which a character in the shifted sequence doesn’t have a consistent partner. We will see that the first alignment that satisfies this requirement, if we examine other alignments in order from less shifted to more, is this one:

```
[593][892][197]n[847][143][542]n[122][527][
                [593][892][197]n[847][143][542]n[122][527][
                ^   ^^   ^^   ^^^   ^^   ^^
```

Each character in the shifted sequence now has a consistent partner in the unshifted one. We can therefore proceed to generate predictions: 5 maps to 8, 4 maps to 2, and 2 maps to 3. Continuing, ] maps to ], and we get the prediction 823], as hoped for.

# ChatGPT Analyzes a Circuit

F

```
User
I've connected a battery to a solenoid. The armature on the solenoid
opens a switch that's between the battery and the solenoid. There's
a spring that normally keeps the switch closed. How will this circuit
behave?
ChatGPT
The circuit you've described forms a basic electrical control system
that includes a solenoid actuator, a switch, and a spring mechanism.
The behavior of this circuit will follow these steps:
Initial State: The spring keeps the switch closed, allowing current
to flow from the battery to the solenoid. This is the system's rest
state, where the spring's force is sufficient to keep the switch
closed despite any resistance or lack thereof in the rest of the
circuit.
Solenoid Activation: When the battery is connected and the switch is
closed (thanks to the spring), current flows through the solenoid.
This current generates a magnetic field within the solenoid, which
creates a force on the armature (a movable iron core or a plunger).
Switch Opening: The movement of the armature, driven by the magnetic
force, acts upon the switch. Given your description, the armature
movement opens the switch. This action breaks the circuit, cutting
off the current flow to the solenoid.
Solenoid Deactivation and Reset: With the circuit open and the current
stopped, the magnetic field collapses, and the force on the armature
disappears. The spring then pushes the switch back to its normally
closed position, ready to start the cycle again.
```





C. Lewis, *Artificial Psychology*, Synthesis Lectures on Human-Centered Informatics,
https://doi.org/10.1007/978-3-031-76646-6

However, there's an inherent contradiction in this setup as described:

If the solenoid only opens the switch and there's no mechanism to keep the switch open independently of the solenoid's action, the system could quickly return to its initial state (with the switch closed) due to the spring's action. This could result in a very brief pulse of current through the solenoid, too brief to perform useful work or even to notice without sensitive instruments.

If the opening of the switch by the solenoid's armature effectively stops the current flow to the solenoid itself, this would suggest a self-interrupting system. Once the switch is opened, the solenoid loses its power, the armature moves back due to the spring (assuming the spring is part of the switch mechanism or there's another mechanism to return the armature), and the switch closes again due to the spring's force, potentially starting the cycle over.

This setup, as described, might not perform a continuous or useful operation without further modifications. For example, adding a delay mechanism, an electronic control system to manage the timing of the switch, or modifying the circuit to allow for a maintained state where the solenoid can hold the switch open for a longer period might be necessary for practical applications.

User

Would this circuit make a noise?

ChatGPT

Yes, the circuit as described would likely produce a noise, especially considering the rapid mechanical movements involved. Here's how the noise could be generated at different stages:

Switch Closing: When the spring forces the switch to close, the contact between the switch parts might produce a clicking sound. The intensity of this sound depends on the switch's design, the force of the spring, and how abruptly the contacts come together.

Solenoid Activation: The activation of the solenoid, which involves the rapid movement of the armature due to the electromagnetic force, can also produce a noticeable sound. This sound is typically a click or a thud, resulting from the armature moving within the solenoid's coil and possibly hitting the end of its travel path.

Switch Opening: The armature's action on the switch, which opens it and interrupts the circuit, could create another clicking sound. This would be similar to the sound of the switch closing but might have a different character depending on how the armature interacts with the switch.

Rapid Cycling: If the circuit operates as a self-interrupting loop, where the solenoid repeatedly opens and closes the switch in quick succession, this could result in a series of rapid clicks or buzzing sounds. The frequency and volume of the noise would depend on how fast the circuit cycles through its open and close states.

The overall noise level would likely be low, comparable to other small electromechanical devices, but it should be audible in a quiet environment. The specific characteristics of the noise (e.g., pitch, volume) would depend on the physical dimensions of the solenoid, the force of the spring, the materials involved, and the design of the switch and armature.

User

Can you think of a use for that?

ChatGPT

Certainly! A circuit that operates in the manner you've described, despite its simplicity, can be adapted for various practical applications. The repetitive action and the noise generated can be utilized in several ways:

Audible Signaling Device: The most straightforward application might be as an audible indicator or alarm. The noise generated from the rapid cycling could serve as a warning or notification mechanism in situations where visual signals might not be sufficient. For example, it could be used in appliances to indicate that a cycle has finished or in safety equipment to warn of a malfunction.

Vibration Generation: The repetitive mechanical action could be used to create vibrations. This could be useful in applications requiring vibration, such as in massage devices, mobile phones, or as a tactile feedback mechanism in electronic interfaces.

Simple Timer or Interrupt Mechanism: By adjusting the spring tension, the solenoid's characteristics, or adding capacitive or resistive elements to the circuit, you could modify the duration between the opening and closing actions of the switch. This could serve as a rudimentary timer for controlling other circuits or devices in a sequence, or as an interrupt mechanism to periodically toggle power or signals in a system.

Educational Tool: This circuit can serve as an excellent educational tool for teaching basic principles of electromagnetism, mechanics, and circuit design. The visual and auditory feedback from the circuit operation makes it a compelling demonstration of these principles in action.

Mechanical Actuation for Small Devices: The repetitive opening and closing action could be harnessed for mechanical actuation in small-scale applications, such as automated pet feeders, where the actuation could trigger the release of small amounts of food at intervals, or in hobbyist projects requiring periodic mechanical movements.

Pest Deterrent: The noise and possible vibration generated could be used as a deterrent for small animals or pests in certain environments, such as gardens or storage areas, where the intermittent noise could discourage pests from settling.

The key to utilizing this circuit effectively lies in creatively adapting its basic operation-repetitive switching caused by the

```
solenoid's actuation and the spring's return force---to suit the needs
of the application. With some modifications and perhaps integration
with other components, the circuit could be tailored to perform a wide
range of useful functions.
```

# GPT on Checking Recall

G

```
If one is trying to recall something one isn't sure of, what would
be a way to check one's accuracy?

ChatGPT

There are several strategies you can employ to check the accuracy of
your recall:

Consult reliable sources: If you're trying to recall factual
information, consulting reliable sources such as textbooks, articles,
or reputable websites can help verify the accuracy of your memory.

Ask others: Sometimes, discussing what you're trying to recall with
others who might have knowledge of the subject can help confirm or
correct your memory.

Cross-referencing: If you have access to multiple sources of
information, cross-referencing what you remember with different
sources can help you identify inconsistencies or confirm accuracy.

Use memory aids: If you've taken notes or written things down
previously, referring back to your notes or records can help you
verify your memory.

Reflect on context: Consider the context in which you encountered the
information you're trying to recall. Sometimes, recalling related
details or events can help you piece together the accuracy of your
memory.

Test yourself: If possible, create a quiz or test for yourself based
on the information you're trying to recall. This can help you identify
gaps in your memory or areas where you may be less confident.

Revisit the information: If time allows, take a break and come back
to the information later. Sometimes, giving your brain time to rest
and process can help clarify your memory.
```





C. Lewis, *Artificial Psychology*, Synthesis Lectures on Human-Centered Informatics,
https://doi.org/10.1007/978-3-031-76646-6

By using one or more of these strategies, you can better assess the accuracy of your recall and improve your overall memory retention.

User

I'm interested in the case where I have ''recalled'' something, but I'm not sure it's right.

ChatGPT

If you're unsure about the accuracy of something you've recalled, there are a few steps you can take to check its accuracy:

Research: Look up the information using reliable sources to verify its correctness. If you're uncertain about a historical fact, scientific concept, or any other type of information, researching it can provide clarity.

Seek clarification: If the information pertains to a specific topic or event, consider reaching out to experts or knowledgeable individuals in that field to confirm or correct your understanding.

Compare with known facts: Cross-reference the recalled information with other facts or knowledge you're confident about. If the recalled information contradicts what you know to be true, it might indicate an error in your memory.

Review related material: Take a broader look at the subject matter or context surrounding the recalled information. Reviewing related material might trigger additional memories or provide additional context to help assess its accuracy.

Reflect on the source: Consider where you initially encountered or learned the information. Reflecting on the reliability and credibility of the source can help you evaluate the likelihood of the recalled information being accurate.

Consider alternative perspectives: Think about alternative interpretations or explanations for the information you've recalled. Considering different viewpoints can help you evaluate its accuracy more objectively.

Test your memory: If possible, create a quiz or test for yourself based on the information you've recalled. Testing your memory can help identify areas of uncertainty or inconsistency that require further investigation.

By employing these strategies, you can better assess the accuracy of the information you've recalled and correct any errors or misconceptions. Remember that it's okay to acknowledge uncertainty and actively seek out clarification when needed.

# Recall Tests

# H

## H.1 With Reminding by User:

User Name all the kinds of brushes you can think of.

ChatGPT

Brushes come in a wide variety of shapes, sizes, and materials, designed for different uses. Here's a list of various kinds of brushes:

Art and Craft Brushes

Paintbrushes (for watercolor, acrylic, oil, gouache, etc.) Foam brushes (for smooth, broad applications) Stencil brushes (for stenciling) Fan brushes (for blending and textural effects) Detail brushes (for fine work) Filbert brushes (oval-shaped for blending) Round brushes (for detail and lines) Flat brushes (for broad strokes) Angle brushes (for curved strokes and filling corners) Mop brushes (for large areas and washes) Rigger brushes (long, thin brushes for lines) Sumi brushes (for East Asian calligraphy and painting)

Hair and Beauty Brushes

Hairbrushes (paddle, round, vented, teasing, etc.) Makeup brushes (foundation, powder, blush, eyeshadow, eyeliner, lip, etc.) Nail brushes (for cleaning and nail art) Shaving brushes (to apply shaving cream or soap)

Cleaning Brushes

Toothbrushes (manual and electric) Scrub brushes (for floors, tiles, and tough stains) Dishwashing brushes (for kitchenware) Bottle brushes (for cleaning the insides of bottles) Toilet brushes (for cleaning toilets) Wire brushes (for removing rust and paint) Pool brushes (for scrubbing pool walls and floors)

Industrial and Professional Brushes



C. Lewis, *Artificial Psychology*, Synthesis Lectures on Human-Centered Informatics, https://doi.org/10.1007/978-3-031-76646-6

Paint rollers (for painting walls and ceilings) Wire wheel brushes (for metalwork) Tube brushes (for cleaning inside tubes and cylinders) Parts cleaning brushes (for machinery and automotive parts) Street sweeper brushes (for cleaning streets) Conveyor brushes (for cleaning conveyor belts) Strip brushes (for sealing and guiding in machinery)

Miscellaneous Brushes

Shoe brushes (for polishing and cleaning shoes) Pet brushes (for grooming) Lint brushes (for removing lint from clothing) Dusting brushes (for furniture and electronics) Barbecue brushes (for cleaning grills) Garden brushes (for cleaning pathways and decking)

This list covers a broad spectrum of brushes, but there are likely more specialized types designed for specific tasks or industries.

User

Does thinking of ''electric drill'' remind you of any other kind of brushes?

ChatGPT

Yes, thinking of an ''electric drill'' does bring to mind another kind of brush: carbon brushes. Carbon brushes are used in many types of electric motors and generators, including those found in electric drills. These brushes conduct electricity between stationary wires and moving parts, typically the rotating shaft or armature. They are made of carbon or a carbon compound, which is chosen for its conductivity, durability, and self-lubricating properties. Carbon brushes play a crucial role in the performance and longevity of electric motors by allowing for an efficient transfer of electric current.

## H.2 With Self-Reminding

I want a list of as many different kinds of brushes as possible. To produce this, begin by listing as many as you can, prompted just by the word ''brushes''. Then think of another word that might remind you of additional kinds that you didn't think of at first. Report the word, and the additional brush kinds. Then think of another word, and repeat for two more reminder words.

ChatGPT

Starting with the prompt ''brushes,'' I'll list as many kinds of brushes as I can think of initially:

Paintbrush Toothbrush Hairbrush Makeup brush Nail brush Toilet brush Dishwashing brush Bottle brush Wire brush Shoe brush Scrub brush Dustpan and brush Broom (as some brooms are essentially large brushes) Shaving brush Electric facial cleansing brush Suede brush Grout brush Pool brush Vegetable brush Floor brush

For the first reminder word, I'll choose ''art.'' This brings to mind additional kinds of brushes related to art or specific applications in painting:

Fan brush Round brush Flat brush Filbert brush Angular brush Detail brush Liner brush Mop brush Sable brush Synthetic brush

For the second reminder word, let's go with ''grooming.'' This word brings to mind additional types of brushes used for personal care or grooming, not previously mentioned:

Beard brush Mustache brush Eyebrow brush Eyelash brush Exfoliating brush Detangling brush Hot air brush Scalp massage brush

For the third reminder word, I'll use ''cleaning.'' This might help us think of more specialized or less common cleaning brushes:

Chimney sweep brush Dryer vent brush Condenser brush (for refrigerators and freezers) Wheel brush (for cars) Gutter brush Pipe cleaner (a type of brush for cleaning pipes) Radiator brush Roof cleaning brush

Lastly, for the fourth reminder word, let's choose ''hobby.'' This can bring to mind brushes used in various hobbies or special interests:

Airbrush (for model painting) Stencil brush Foam brush (often used for crafts) Pottery brush (specific shapes and stiffness for pottery) Watercolor brush Calligraphy brush Modelling brush (for miniatures and models) Leather dye brush

This approach helps in uncovering a wide variety of brushes by associating them with different contexts or uses prompted by the reminder words.

# Glenberg and Robertson Test Items

Experiment 1:

Mike was freezing while walking up State Street into a brisk wind. He knew that he had to get his face covered pretty soon or he would get frostbite. Unfortunately, he didn't have enough money to buy a scarf. Which of the following two sentences makes more sense?

(1) Being clever, he walked into a store and bought a newspaper to cover his face.
(2) Being clever, he walked into a store and bought a matchbook to cover his face.

Marissa forgot to bring her pillow on her camping trip. Which of the following two sentences makes more sense?

(1) As a substitute for her pillow, she filled up an old sweater with leaves.
(2) As a substitute for her pillow, she filled up an old sweater with water.

Experiment 2:

Kate was cleaning her kitchen on Sunday morning after a big party she had the night before. How much sense does the following sentence make? Use a scale of 1 (virtual nonsense) to 7 (completely sensible).

Since she couldn't reach the ceiling, she stuck her broom up in the air to try to get a piece of gum off her ceiling tile.

Kate was cleaning her kitchen on Sunday morning after a big party she had the night before. How much sense does the following sentence make? Use a scale of 1 (virtual nonsense) to 7 (completely sensible).

She got down on her hands and knees to scrape the beer stains off the ceiling tile.



C. Lewis, *Artificial Psychology*, Synthesis Lectures on Human-Centered Informatics, https://doi.org/10.1007/978-3-031-76646-6

Kate was cleaning her kitchen on Sunday morning after a big party she had the night before. How much sense does the following sentence make? Use a scale of 1 (virtual nonsense) to 7 (completely sensible).

She got down on her hands and knees to scrape the beer stains off the floor tile.

Kate was cleaning her kitchen on Sunday morning after a big party she had the night before. How much sense does the following sentence make? Use a scale of 1 (virtual nonsense) to 7 (completely sensible).

Since she couldn't reach the ceiling, she stuck her broom up in the air to try to get a piece of gum off her floor tile.

Experiment 3:

Read this paragraph: Kenny sat in the tree house and patiently waited. He clutched the jar of green ooze in his hand, and watched the approaching school bus move closer to his house. The teenage girl stepped off and walked towards the tree house unaware of the little boy above her taking the cap off the jar. Kenny waited until she was directly beneath him, and an evil grin spread across his face. Then, Kenny slimed his sister.

In the context of the paragraph, how much sense does the last sentence make? Use a scale from 1 (virtual nonsense) to 7 (completely sensible).

Read this paragraph: Lori loved her new table, until she noticed that everything she placed on it slid off to the left. The left back leg was lower than all the others. She could not imagine how to fix the slant. Then she spotted a pile of hard-covered books in the corner. She booked the leg.

In the context of the paragraph, how much sense does the last sentence make? Use a scale from 1 (virtual nonsense) to 7 (completely sensible).

Read this paragraph: Lori was having a really bad day. She could not find her textbook and she was late for class. Frantically, she ran over to the table where there was a pile of books. On the way, she banged her leg on the chair. She booked the leg.

In the context of the paragraph, how much sense does the last sentence make? Use a scale from 1 (virtual nonsense) to 7 (completely sensible).

Read this paragraph: Sebastian was perusing the latest issue of Newsweek when he was disturbed by a most annoying buzzing noise. He looked around the room to determine the source of this disturbance, and saw that a fly was patrolling the vicinity. Its incessant buzzing was making Sebastian insane. He had no choice but to terminate with extreme prejudice. So, he rolled up his Newsweek and waited patiently. When the fly came to rest on the coffee table in front of Sebastian, he recognized his opportunity. He magazined it.

In the context of the paragraph, how much sense does the last sentence make? Use a scale from 1 (virtual nonsense) to 7 (completely sensible).

Read this paragraph: Sebastian was perusing the latest issue of Newsweek. He became disturbed as he read an article about rising rates of home invasions in his vicinity. Sebastian decided to follow the advice of a security expert quoted in the magazine by purchasing a home security alarm. The salesman at the electronics store thought Sebastian was insane when he insisted on having the alarm installed that very day, but agreed when Sebastian threatened to terminate the sale. The alarm woke Sebastian when it began buzzing one evening. He recognized his opportunity. He magazined it.

In the context of the paragraph, how much sense does the last sentence make? Use a scale from 1 (virtual nonsense) to 7 (completely sensible).

# Example Items from Glenberg and Robertson, with Responses from ChatGPT

J

## J.1 Items from Experiment 1, with Responses from ChatGPT

(1.1) Mike was freezing while walking up State Street into a brisk wind. He knew that he had to get his face covered pretty soon or he would get frostbite. Unfortunately, he didn't have enough money to buy a scarf.

Which of the following two sentences makes more sense?

(1) Being clever, he walked into a store and bought a newspaper to cover his face.
(2) Being clever, he walked into a store and bought a matchbook to cover his face.

ChatGPT

The sentence that makes more sense in this context is:

(1) Being clever, he walked into a store and bought a newspaper to cover his face.

This choice is more logical because a newspaper is large enough to cover the face and provide some protection against the wind and cold. A matchbook, on the other hand, is too small to effectively cover and protect the face from frostbite.

(1.2) Marissa forgot to bring her pillow on her camping trip. Which of the following two sentences makes more sense?

(1) As a substitute for her pillow, she filled up an old sweater with leaves.
(2) As a substitute for her pillow, she filled up an old sweater with water.





C. Lewis, *Artificial Psychology*, Synthesis Lectures on Human-Centered Informatics,
https://doi.org/10.1007/978-3-031-76646-6

```
ChatGPT

The sentence that makes more sense in this context is:

(1) As a substitute for her pillow, she filled up an old sweater with
leaves.

Filling an old sweater with leaves is a plausible and practical
solution when one forgets a pillow on a camping trip. Leaves can
provide a cushioned, albeit temporary, substitute for a pillow. On
the other hand, filling a sweater with water would result in a wet,
uncomfortable, and impractical object to rest one's head on. Therefore,
the first sentence is more logical and appropriate in this scenario.
```

## J.2 Items from Experiment 2, with ChatGPT Responses

The context for all these items was:

```
Kate was cleaning her kitchen on Sunday morning after a big party she
had the night before. How much sense does the following sentence make?
Use a scale of 1 (virtual nonsense) to 7 (completely sensible).

(2.1) Since she couldn't reach the ceiling, she stuck her broom up
in the air to try to get a piece of gum off her ceiling tile.

ChatGPT:

The sentence ``Since she couldn't reach the ceiling, she stuck her
broom up in the air to try to get a piece of gum off her ceiling tile''
makes a fair amount of sense in the context provided. It is a logical
action to take if someone is unable to reach the ceiling by hand and
there's something like gum stuck there. I would rate it a 6 on a scale
from 1 (virtual nonsense) to 7 (completely sensible). This rating
reflects that the action is practical and understandable, though
slightly unconventional.

(2.2) She got down on her hands and knees to scrape the beer stains
off the ceiling tile.

ChatGPT:

The sentence ``She got down on her hands and knees to scrape the beer
stains off the ceiling tile'' would be rated as a 2 on the scale of
sensibility (where 1 is virtually nonsensical and 7 is completely
sensible). The action of getting ``down on her hands and knees'' is
typically associated with cleaning floors or low surfaces, not
ceilings. Therefore, the posture described is illogical for the task
of cleaning a ceiling. This creates a contradiction in the sentence,
rendering it mostly nonsensical in a practical context.

(2.3) She got down on her hands and knees to scrape the beer stains
off the floor tile.

ChatGPT:
```

The sentence ''She got down on her hands and knees to scrape the beer stains off the floor tile'' would rate a 7 on the scale of 1 (virtual nonsense) to 7 (completely sensible). It makes complete sense in the context of cleaning up after a party, as it describes a typical action someone might take to remove tough stains from a floor.

(2.4) Since she couldn't reach the ceiling, she stuck her broom up in the air to try to get a piece of gum off her floor tile.

ChatGPT:

The sentence you provided would rate as a 2 on the scale of 1 (virtual nonsense) to 7 (completely sensible). It contains elements that are somewhat logical, such as cleaning after a party and encountering a piece of gum on the floor. However, the action of using a broom to reach the ceiling in an attempt to remove gum from a floor tile is illogical and contradictory, significantly reducing the sentence's overall sensibility.

## J.3 Items from Experiment 3, with Responses from ChatGPT

(3.1) Read this paragraph: Kenny sat in the tree house and patiently waited. He clutched the jar of green ooze in his hand, and watched the approaching school bus move closer to his house. The teenage girl stepped off and walked towards the tree house unaware of the little boy above her taking the cap off the jar. Kenny waited until she was directly beneath him, and an evil grin spread across his face. Then, Kenny slimed his sister.

In the context of the paragraph, how much sense does the last sentence make? Use a scale from 1 (virtual nonsense) to 7 (completely sensible).

ChatGPT:

The last sentence of the paragraph, ''Then, Kenny slimed his sister'', makes complete sense in the context provided and can be rated as a 7 on a scale from 1 to 7. The narrative sets up a scenario where Kenny is waiting in a tree house with a jar of green ooze as he watches a school bus approach. A teenage girl, who is later identified as his sister, steps off the bus and walks towards the tree house, unaware of Kenny's intentions. The culmination of these events in Kenny sliming his sister is a logical and sensible conclusion to the sequence of actions described in the paragraph.

(3.2) Read this paragraph: Lori loved her new table, until she noticed that everything she placed on it slid off to the left. The left back leg was lower than all the others. She could not imagine how to fix the slant. Then she spotted a pile of hard-covered books in the corner. She booked the leg.

In the context of the paragraph, how much sense does the last sentence make? Use a scale from 1 (virtual nonsense) to 7 (completely sensible).

ChatGPT:

The last sentence of the paragraph makes a lot of sense in the given context and would rate a 7 on the scale from 1 (virtual nonsense) to 7 (completely sensible). The phrase ''she booked the leg'' is a creative way of saying that Lori used books to support the lower leg of the table to correct its slant. This is a practical and common solution for stabilizing a wobbly table, making the sentence both logical and relevant to the situation described.

(3.3) Read this paragraph: Lori was having a really bad day. She could not find her textbook and she was late for class. Frantically, she ran over to the table where there was a pile of books. On the way, she banged her leg on the chair. She booked the leg.

In the context of the paragraph, how much sense does the last sentence make? Use a scale from 1 (virtual nonsense) to 7 (completely sensible).

ChatGPT:

The last sentence, ''She booked the leg'', in the provided paragraph does not make much sense in the given context. The term ''booked'' does not fit logically with the action described or the preceding narrative. Therefore, on a scale from 1 (virtual nonsense) to 7 (completely sensible), this sentence would be rated a 1. It appears to be a typographical or semantic error, possibly intended to be ''bumped'' or ''bruised'' her leg, which would be more coherent with the context of running and banging her leg on the chair.

(3.4) Read this paragraph: Sebastian was perusing the latest issue of Newsweek when he was disturbed by a most annoying buzzing noise. He looked around the room to determine the source of this disturbance, and saw that a fly was patrolling the vicinity. It's incessant buzzing was making Sebastian insane. He had no choice but to terminate with extreme prejudice. So, he rolled up his Newsweek and waited patiently. When the fly came to rest on the coffee table in front of Sebastian, he recognized his opportunity. He magazined it.

In the context of the paragraph, how much sense does the last sentence make? Use a scale from 1 (virtual nonsense) to 7 (completely sensible).

ChatGPT:

The last sentence in the paragraph, ''He magazined it'', is somewhat unconventional but still makes sense in context. The use of ''magazined'' as a verb is non-standard, but the narrative makes it clear that Sebastian used the magazine to swat the fly. The context of him rolling up the magazine and waiting for the right moment to act provides enough information for the reader to understand the intended meaning, even though the verb form is creatively coined. On a scale from 1 (virtual nonsense) to 7 (completely sensible), I would rate this sentence a 5. It's not standard English, but it conveys the intended action effectively in the context of the story.

(3.5) Read this paragraph: Sebastian was perusing the latest issue of Newsweek. He became disturbed as he read an article about rising rates of home invasions in his vicinity. Sebastian decided to follow the advice of a security expert quoted in the magazine by purchasing a home security alarm. The salesman at the electronics store thought Sebastian was insane when he insisted on having the alarm installed that very day, but agreed when Sebastian threatened to terminate the sale. The alarm woke Sebastian when it began buzzing one evening. He recognized his opportunity. He magazined it.

In the context of the paragraph, how much sense does the last sentence make? Use a scale from 1 (virtual nonsense) to 7 (completely sensible).

The last sentence, ''He recognized his opportunity. He magazined it'', is quite ambiguous and unclear in its meaning, particularly the phrase ''he magazined it''. This phrase does not conform to standard uses of English, as ''magazine'' is not typically used as a verb, and its intended meaning in this context is not clear. It could be interpreted as an attempt to relate back to his reading of Newsweek magazine, perhaps implying he applied knowledge or tactics from the magazine. However, this interpretation requires a significant assumption.

Given the lack of clarity and unconventional usage of ''magazined'', I would rate this sentence a 2 on a scale of 1 (virtual nonsense) to 7 (completely sensible). It has some connection to the rest of the paragraph through the mention of the magazine, but the actual action described (''magazined it'') is not sensibly conveyed.

# Code for Sawtooth Curves

Prompt:

I need a Web application, on a single page, with Javascript, that draws a 45 °C right triangle, fairly large. The legs of the triangle should lie along the x- and y-axes. Then it draws a family of five sawtooth curves, like this. All the curves start where the hypoteneuse hits the x-axis. Call the side of the triangle s. The first curve goes s/2 to the left, then s/2 up, then s/2 left. The second curve goes s/4 to the left, then s/4 up, then s/4 left, and so on, until it hits the y-axis. The third curve does the same thing, only moving s/8 each time, until it hits the y-axis. The last two curves do the same thing, only with steps s/16 and s/32. Put everything in a single html file, including the javascript.

GPT Response:

To create a web application that draws a 45° right triangle along with a family of five sawtooth curves as described, you can follow the structure provided below. This solution uses HTML for the structure, and JavaScript along with the Canvas API to draw the shapes.

Here's a simple template for your requirements. Please note, for clarity and simplicity, this example assumes the side of the triangle (s) is a fixed length. You may adjust s to change the size of the triangle and the curves.

```
The code:
<!DOCTYPE html>
<html lang="en">
<head>
<meta charset="UTF-8">
<title>Triangle and Sawtooth Curves</title>
<style>
  body {
    display: flex;
    justify-content: center;
    align-items: center;
```



C. Lewis, *Artificial Psychology*, Synthesis Lectures on Human-Centered Informatics,
https://doi.org/10.1007/978-3-031-76646-6

```
    height: 100vh;
  }
  canvas {
    border: 1px solid black;
  }
</style>
</head>
<body>
<canvas id="canvas"></canvas>
<script>
  const canvas = document.getElementById('canvas');
  const ctx = canvas.getContext('2d');

  // Set canvas size and the side of the triangle
  const s = 300; // Side length of the triangle
  canvas.width = s;
  canvas.height = s;

  // Function to draw the right triangle
  function drawTriangle(sideLength) {
    ctx.beginPath();
    ctx.moveTo(0, sideLength);
    ctx.lineTo(0, 0);
    ctx.lineTo(sideLength, sideLength);
    ctx.closePath();
    ctx.stroke();
  }

  // Function to draw sawtooth curves
  function drawSawtooth(startX, startY, step, sideLength) {
    let x = startX;
    let y = startY;

    while (y > 0) {
      ctx.lineTo(x - step, y);
      ctx.lineTo(x - step, y - step);
      x -= step;
      y -= step;
    }
    ctx.stroke();
  }

  // Drawing the triangle
  drawTriangle(s);
```

```
    // Drawing sawtooth curves
    ctx.beginPath();
    ctx.moveTo(s, s);
    for (let i = 0; i < 5; i++) {
      drawSawtooth(s, s, s / Math.pow(2, i+1), s);
      ctx.moveTo(s, s);
    }
  </script>
  </body>
  </html>
```

More discussion from chatGPT: This code does the following:

Sets up a basic HTML5 document with a canvas element to draw on. Defines CSS to center the canvas in the viewport and adds a border for visibility. Includes JavaScript to: Define the canvas size and the side length (s) of the triangle. Draw a 45° right triangle using the drawTriangle function. Draw a family of five sawtooth curves starting from the bottom-right corner of the triangle towards the y-axis, with each subsequent curve having steps half the size of the previous one, as specified in the drawSawtooth function. Iteratively calls drawSawtooth for each of the five curves with decreasing step sizes, starting from s/2 and halving each time. You can copy and paste this code into an HTML file and open it in a web browser to see the drawing. Adjust the value of s to change the size of the triangle and curves to fit your needs.

# References

Akyürek, A. F., Akyürek, E., Choshen, L., Wijaya, D., & Andreas, J. (2024). Deductive closure training of language models for coherence, accuracy, and updatability. *arXiv preprint* arXiv:2401.08574.
Amin, T. G., & Levrini, O. (2018). Converging perspectives on conceptual change. *Mapping an emerging paradigm in the learning sciences.*
Anderson, J. R. (1976). *Language, memory, and thought.* Lawrence Erlbaum Associates.
Anderson, J. R. (1983). *The architecture of cognition.* Harvard University Press.
Anderson, J. R., & Schunn, C. D. (2013). Implications of the act-r learning theory: No magic bullets. In *Advances in instructional psychology, volume 5* (pp. 1–33). Routledge.
Barrett, L. F. (2017). *How emotions are made: The secret life of the brain.* Houghton Mifflin Harcourt.
Bartlett, F. C. (1932). *Remembering: A study in experimental and social psychology.* Cambridge university press.
Baverstock, A. (2015). Last night was a real buzz! the shocking drinking game seeing a surge in mexican bars - where party-goers are electrocuted until they scream [Online; accessed 26-March 2024].
Benacerraf, P. (1965). What numbers could not be. *The philosophical review, 74*(1), 47–73.
Berglund, L., Tong, M., Kaufmann, M., Balesni, M., Stickland, A. C., Korbak, T., & Evans, O. (2023). The reversal curse: Llms trained on" a is b" fail to learn" b is a". *arXiv preprint* arXiv:2309.12288.
Biber, D., & Gray, B. (2011). Grammatical change in the noun phrase: The influence of written language use. *English Language & Linguistics, 15*(2), 223–250.
Bickel, J. (2020). Multiple realizability. In E. N. Zalta (Ed.), *The stanford encyclopedia of philosophy* (Summer 2020 ed.). https://plato.stanford.edu/archives/sum2020/entries/multiple-realizability/
Blackwell, A. (2024). Moral codes [Online; accessed 28-March 2024].
Blakeslee, S. (2006). Cells that read minds. *New York Times, 10*(1).
Bovair, S., Kieras, D. E., & Polson, P. G. (1990). The acquisition and performance of text-editing skill: A cognitive complexity analysis. *Human-Computer Interaction, 5*(1), 1–48.
Bracher, M. (2006). *Radical pedagogy: Identity, generativity, and social transformation.* Springer.
Brandom, R. B. (2001). *Articulating reasons: An introduction to inferentialism.* Harvard University Press.



C. Lewis, *Artificial Psychology*, Synthesis Lectures on Human-Centered Informatics,
https://doi.org/10.1007/978-3-031-76646-6

Brooks, R. (1986). A robust layered control system for a mobile robot. *IEEE journal on robotics and automation*, *2*(1), 14–23.

Carey, S. (2009). *The origin of concepts*. Oxford University Press, New York.

Carpenter, M., Nagell, K., Tomasello, M., Butterworth, G., & Moore, C. (1998). Social cognition, joint attention, and communicative competence from 9 to 15 months of age. *Monographs of the society for research in child development*, i–174.

Carraher, T. N., Carraher, D. W., & Schliemann, A. D. (1985). Mathematics in the streets and in schools. *British journal of developmental psychology*, *3*(1), 21–29.

Catmur, C., Gillmeister, H., Bird, G., Liepelt, R., Brass, M., & Heyes, C. (2008). Through the looking glass: Counter-mirror activation following incompatible sensorimotor learning. *European Journal of Neuroscience*, *28*(6), 1208–1215.

Charniak, E. (1968). *Carps, a program which solves calculus word problems* (tech. rep.). Project MAC, MIT.

Chittka, L., Dyer, A. G., Bock, F., & Dornhaus, A. (2003). Bees trade off foraging speed for accuracy. *Nature*, *424*(6947), 388–388.

Chomsky, N. (1980a). *Rules and representations*. Columbia University Press.

Chomsky, N. (1980b). Rules and representations. *Behavioral and brain sciences*, *3*(1), 1–15.

Chomsky, N., Roberts, I., & Watumull, J. (2023). Noam chomsky: The false promise of chatgpt. *The New York Times*, *8*.

Clark, A. (2013). Whatever next? predictive brains, situated agents, and the future of cognitive science. *Behavioral and brain sciences*, *36*(3), 181–204.

Clark, A. (2023). *The experience machine: How our minds predict and shape reality*. Pantheon.

Cohen, N. J., & Squire, L. R. (1981). Retrograde amnesia and remote memory impairment. *Neuropsychologia*, *19*(3), 337–356.

Craik, F. I., & Tulving, E. (1975). Depth of processing and the retention of words in episodic memory. *Journal of experimental Psychology: general*, *104*(3), 268.

Devlin, J., Chang, M.-W., Lee, K., & Toutanova, K. (2018). Bert: Pre-training of deep bidirectional transformers for language understanding. *arXiv preprint* arXiv:1810.04805.

DiSessa, A. A. (1993). Toward an epistemology of physics. *Cognition and instruction*, *10*(2-3), 105–225.

DiSessa, A. A. (2018). A friendly introduction to "knowledge in pieces": Modeling types of knowledge and their roles in learning. *Invited lectures from the 13th international congress on mathematical education*, 65–84.

Driess, D., Xia, F., Sajjadi, M. S., Lynch, C., Chowdhery, A., Ichter, B., Wahid, A., Tompson, J., Vuong, Q., Yu, T., et al. (2023). Palm-e: An embodied multimodal language model. *arXiv preprint* arXiv:2303.03378.

Elhage, N., Nanda, N., Olsson, C., Henighan, T., Joseph, N., Mann, B., Askell, A., Bai, Y., Chen, A., Conerly, T., et al. (2021). A mathematical framework for transformer circuits. *Transformer Circuits Thread*, *1*, 1.

Elman, J. L. (1990). Finding structure in time. *Cognitive science*, *14*(2), 179–211.

Ericsson, K. A., & Chase, W. G. (1982). Exceptional memory: Extraordinary feats of memory can be matched or surpassed by people with average memories that have been improved by training. *American Scientist*, *70*(6), 607–615.

Evans, N., & Levinson, S. C. (2009). The myth of language universals: Language diversity and its importance for cognitive science. *Behavioral and brain sciences*, *32*(5), 429–448.

Eyzaguirre, C., del Rio, F., Araujo, V., & Soto, A. (2021). Dact-bert: Differentiable adaptive computation time for an efficient bert inference. *arXiv preprint* arXiv:2109.11745.

Fauconnier, G., & Turner, M. (2003). Conceptual blending, form and meaning. *Recherches en communication*, *19*, 57–86.

Fernyhough, C. (2008). Getting vygotskian about theory of mind: Mediation, dialogue, and the development of social understanding. *Developmental review*, *28*(2), 225–262.

Flash, T., & Hochner, B. (2005). Motor primitives in vertebrates and invertebrates. *Current opinion in neurobiology*, *15*(6), 660–666.

Flynn, A. M., Brooks, R. A., Wells III, W. M., & Barrett, D. S. (1989). Intelligence for miniature robots. *sensors and actuators*, *20*(1-2), 187–196.

Forbus, K. D. (1988). Qualitative physics: Past, present, and future. In *Exploring artificial intelligence* (pp. 239–296). Elsevier.

Frankish, K. (2016). Illusionism as a theory of consciousness. *Journal of Consciousness Studies*, *23*(11-12), 11–39.

Garfinkel, H. (1969). *Studies in ethnomethodology*. Prentice-Hall.

Gentner, D. (1983). Structure-mapping: A theoretical framework for analogy. *Cognitive science*, *7*(2), 155–170.

Gentner, D. (2010). Bootstrapping the mind: Analogical processes and symbol systems. *Cognitive science*, *34*(5), 752–775.

Geva, M., Schuster, R., Berant, J., & Levy, O. (2020). Transformer feed-forward layers are key-value memories. *arXiv preprint* arXiv:2012.14913.

Gilbers, D., & De Hoop, H. (1998). Conflicting constraints: An introduction to optimality theory. *Lingua*, *104*(1-2), 1–12.

Gilligan, J. (2000). Punishment and violence: Is the criminal law based on one huge mistake? *Social Research*, 745–772.

Glenberg, A. M., & Robertson, D. A. (2000). Symbol grounding and meaning: A comparison of high-dimensional and embodied theories of meaning. *Journal of memory and language*, *43*(3), 379–401.

Gold, E. M. (1967). Language identification in the limit. *Information and control*, *10*(5), 447–474.

Goldin-Meadow, S., Kim, S., & Singer, M. (1999). What the teacher's hands tell the student's mind about math. *Journal of educational psychology*, *91*(4), 720.

Haidt, J. (2001). The emotional dog and its rational tail: A social intuitionist approach to moral judgment. *Psychological review*, *108*(4), 814.

Harman, G. (1986). *Change in view: Principles of reasoning*. The MIT Press.

Harnad, S. (1990). The symbol grounding problem. *Physica D: Nonlinear Phenomena*, *42*(1-3), 335–346.

Heitz, R. P. (2014). The speed-accuracy tradeoff: History, physiology, methodology, and behavior. *Frontiers in neuroscience*, *8*, 86875.

Heritage, J. (2013). *Garfinkel and ethnomethodology*. John Wiley & Sons.

Heyes, C., & Catmur, C. (2022). What happened to mirror neurons? *Perspectives on Psychological Science*, *17*(1), 153–168.

Hofstadter, D. R., Mitchell, M., et al. (1995). The copycat project: A model of mental fluidity and analogy-making. *Advances in connectionist and neural computation theory*, *2*, 205–267.

Hofstadter, D. R., & Sander, E. (2013). *Surfaces and essences: Analogy as the fuel and fire of thinking*. Basic books.

Howell, W. C., & Kreidler, D. L. (1963). Information processing under contradictory instructional sets. *Journal of Experimental Psychology*, *65*(1), 39.

Huemer, M. (2005). *Ethical intuitionism*. Palgrave MacMillan.

Huff, I. (2022). Qanon beliefs have increased since 2021 as americans are less likely to reject conspiracies. *Public Religion Research Institute*, *6*, 2022.

James, W. (1890). *Principles of psychology*. Dover.

Kahneman, D. (2011). *Thinking, fast and slow*. macmillan.

Keil, F. C. (1992). *Concepts, kinds, and cognitive development*. mit Press.

Kim, J. (1997). Supervenience, emergence, and realization in the philosophy of mind. *Mindscapes: Philosophy, science, and the mind*, *5*, 271.
Kintsch, W. (1974). *The representation of meaning in memory*. Lawrence Erlbaum Associates.
Kolers, P. A. (1975). Specificity of operations in sentence recognition. *Cognitive Psychology*, *7*(3), 289–306.
Kosinski, M. (2023). Evaluating large language models in theory of mind tasks. *arXiv e-prints*, arXiv–2302.
Lahlou, S. (2018). *Installation theory: The societal construction and regulation of behaviour*. Cambridge University Press.
Lakoff, G., & Johnson, M. (2008). *Metaphors we live by*. University of Chicago press.
Lambert, N., Castricato, L., von Werra, L., & Havrilla, A. (2022). Illustrating reinforcement learning from human feedback (rlhf) [Online; accessed 26-March 2024].
Leinster, T. (2014). Rethinking set theory. *The American Mathematical Monthly*, *121*(5), 403–415.
Lewis, C. (2023). Large language models and the psychology of programming. In D. Brown & S. Green (Eds.), *Ppig 2023 proceedings* (pp. 77–95).
Lewis, C. H., & Anderson, J. R. (1976). Interference with real world knowledge. *Cognitive Psychology*, *8*(3), 311–335.
Lohmar, D. (2017). The unconscious and the non-linguistic mode of thinking. In D. Legrand & D. Trigg (Eds.), *Unconsciousness between phenomenology and psychoanalysis, contributions to phenomenology 88*. Springer.
Lovett, M. C., & Anderson, J. R. (2005). Thinking as a production system. *The Cambridge handbook of thinking and reasoning*, 401–429.
Luchins, A. S. (1942). Mechanization in problem solving: The effect of einstellung. *Psychological monographs*, *54*(6), i.
Maass, W. (2014). Noise as a resource for computation and learning in networks of spiking neurons. *Proceedings of the IEEE*, *102*(5), 860–880.
Mahowald, K., Ivanova, A. A., Blank, I. A., Kanwisher, N., Tenenbaum, J. B., & Fedorenko, E. (2023). Dissociating language and thought in large language models: A cognitive perspective. *arXiv preprint* arXiv:2301.06627.
Mandler, J. M., & Mandler, G. (1964). Thinking: From association to gestalt.
Mayne, A. (n.d.). Overview of the openai api & gpt-3 [Online; accessed 26-March 2024].
Miller, G. A. (1956). The magical number seven, plus or minus two: Some limits on our capacity for processing information. *Psychological review*, *63*(2), 81.
Minstrell, J. (1982). Explaining the "at rest" condition of an object. *The physics teacher*, *20*(1), 10–14.
Mitchell, M. (2020a). Can gpt-3 make analogies? [Online; accessed 18-March 2024].
Mitchell, M. (2020b). Follow-up to "can gpt-3 make analogies?" [Online; accessed 18-March 2024].
Morin, A. (2009). Inner speech and consciousness. In W. P. Banks (Ed.), *Encyclopedia of consciousness*. Elsevier.
Newell, A., & Simon, H. A. (2007). Computer science as empirical inquiry: Symbols and search. In *Acm turing award lectures* (p. 1975).
Nisbett, R. E., & Wilson, T. D. (1977). Telling more than we can know: Verbal reports on mental processes. *Psychological review*, *84*(3), 231.
Norman, D. A., Rumelhart, D. E., & the LNR Research Group. (1975). *Explorations in cognition*. Freeman.
Olsson, C., Elhage, N., Nanda, N., Joseph, N., DasSarma, N., Henighan, T., Mann, B., Askell, A., Bai, Y., Chen, A., et al. (2022). In-context learning and induction heads. *arXiv preprint* arXiv:2209.11895.
Parisi, G. I., Kemker, R., Part, J. L., Kanan, C., & Wermter, S. (2019). Continual lifelong learning with neural networks: A review. *Neural networks*, *113*, 54–71.

Parr, T., Pezzulo, G., & Friston, K. J. (2022). *Active inference: The free energy principle in mind, brain, and behavior*. MIT Press.

Pavlick, E. (2023). Symbols and grounding in large language models. *Philosophical Transactions of the Royal Society A, 381*(2251), 20220041.

Pinker, S. (2013). *Learnability and cognition, new edition: The acquisition of argument structure*. MIT press.

Polson, P. G., Muncher, E., & Engelbeck, G. (1986). A test of a common elements theory of transfer. *ACM SIGCHI Bulletin, 17*(4), 78–83.

Pribram, K. H. (1981). The brain, the telephone, the thermostat, the computer, and the hologram. *Cognition and Brain Theory, 4*(2), 105–122.

Rieber, R. W., & Carton, A. S. (1987). *The collected works of ls vygotsky: Problems of general psychology, including the volume thinking and speech*. Springer.

Riseborough, M. G. (1981). Physiographic gestures as decoding facilitators: Three experiments exploring a neglected facet of communication. *Journal of Nonverbal Behavior, 5*, 172–183.

Ritter, F. E., Tehranchi, F., & Oury, J. D. (2019). Act-r: A cognitive architecture for modeling cognition. *Wiley Interdisciplinary Reviews: Cognitive Science, 10*(3), e1488.

Rumelhart, D. E., & Abrahamson, A. A. (1973). A model for analogical reasoning. *Cognitive Psychology, 5*(1), 1–28.

Rupert, R. D. (2009). *Cognitive systems and the extended mind*. Oxford University Press.

Ryle, G. (2009). The concept of mind. Routledge.

Sauter, D. A., Eisner, F., Ekman, P., & Scott, S. K. (2010). Cross-cultural recognition of basic emotions through nonverbal emotional vocalizations. *Proceedings of the National Academy of Sciences, 107*(6), 2408–2412.

Sauter, D. A., Eisner, F., Ekman, P., & Scott, S. K. (2015). Emotional vocalizations are recognized across cultures regardless of the valence of distractors. *Psychological science, 26*(3), 354–356.

Schiffer, S. (2003). *The things we mean*. Clarendon Press.

Schlag, I., Smolensky, P., Fernandez, R., Jojic, N., Schmidhuber, J., & Gao, J. (2019). Enhancing the transformer with explicit relational encoding for math problem solving. *arXiv preprint* arXiv:1910.06611.

Searle, J. R. (1982). The chinese room revisited. *Behavioral and brain sciences, 5*(2), 345–348.

Shane, J. (2017). The yoneda embedding [Online; accessed 26-March 2024].

Shane, J. (2022). Interview with a squirrel [Online; accessed 26-March 2024].

Simonyi, C., Christerson, M., & Clifford, S. (2006). Intentional software. *Proceedings of the 21st annual ACM SIGPLAN conference on Object-oriented programming systems, languages, and applications*, 451–464.

Smolensky, P., McCoy, R. T., Fernandez, R., Goldrick, M., & Gao, J. (2022). Neurocompositional computing in human and machine intelligence: A tutorial. *Microsoft Technical Report MSR-TR-2022*.

Stich, S. P. (1978). Autonomous psychology and the belief-desire thesis. *Monist, 64*(4), 573–591.

Stich, S. P. (1983). *From folk psychology to cognitive science*. MIT/Bradford Press, Cambridge.

Stocco, A., Rice, P., Thomson, R., Smith, B., Morrison, D., & Lebiere, C. (2024). An integrated computational framework for the neurobiology of memory based on the act-r declarative memory system. *Computational Brain & Behavior, 7*(1), 129–149.

Strasser, C., & Antonelli, G. A. (2024). Non-monotonic logic. In U. Zalta Edward.N. & Nodelman (Ed.), *The stanford encyclopedia of philosophy* (Summer 2024 ed.). https://plato.stanford.edu/archives/sum2024/entries/logic-nonmonotonic/

Suhay, E. (2015). Explaining group influence: The role of identity and emotion in political conformity and polarization. *Political Behavior, 37*, 221–251.

Taatgen, N. A., Huss, D., Dickison, D., & Anderson, J. R. (2008). The acquisition of robust and flexible cognitive skills. *Journal of Experimental Psychology: General, 137*(3), 548.

Tenney, I., Das, D., & Pavlick, E. (2019). Bert rediscovers the classical nlp pipeline. *arXiv preprint* arXiv:1905.05950.

Thissen-Roe, A., Hunt, E., & Minstrell, J. (2004). The diagnoser project: Combining assessment and learning. *Behavior research methods, instruments, & computers, 36*(2), 234–240.

Thucydides, M., et al. (2013). The war of the peloponnesians and the athenians. *(No Title)*.

Tomasello, M. (2019). *Becoming human: A theory of ontogeny*. Harvard University Press.

Tomasello, M., Carpenter, M., & Hobson, R. P. (2005). The emergence of social cognition in three young chimpanzees. *Monographs of the Society for Research in Child development*, i–152.

Tulving, E. (1972). Episodic and semantic memory. In E. Tulving & W. Donaldson (Eds.), *Organization of memory* (pp. 381–403). Academic Press.

Turner, M. (2020). Constructions and creativity. *Cognitive Semiotics, 13*(1), 20202019.

Undorf, M., & Bröder, A. (2020). Cue integration in metamemory judgements is strategic. *Quarterly Journal of Experimental Psychology, 73*(4), 629–642.

University of Oklahoma, I. o. G. R., & Sherif, M. (1961). *Intergroup conflict and cooperation: The robbers cave experiment* (Vol. 10). University Book Exchange Norman, OK.

Van Dijk, T. A., Kintsch, W., et al. (1983). Strategies of discourse comprehension.

Veres, C., & Sampson, J. (2023). Self supervised learning and the poverty of the stimulus. *Data & Knowledge Engineering, 147*, 102208.

Von Helmholtz, H. (2013). *Treatise on physiological optics, volume iii* (Vol. 3). Courier Corporation.

Vong, W. K., Wang, W., Orhan, A. E., & Lake, B. M. (2024). Grounded language acquisition through the eyes and ears of a single child. *Science, 383*(6682), 504–511.

Wang, X., Wei, J., Schuurmans, D., Le, Q., Chi, E., Narang, S., Chowdhery, A., & Zhou, D. (2022). Self-consistency improves chain of thought reasoning in language models. *arXiv preprint* arXiv:2203.11171.

Waterman, D. A. (1975). Adaptive production systems. *IJCAI, 4*, 296–303.

Watson, J. B. (1913). Psychology as the behaviorist views it. *Psychological review, 20*(2), 158.

Webb, T., Holyoak, K. J., & Lu, H. (2023). Emergent analogical reasoning in large language models. *Nature Human Behaviour, 7*(9), 1526–1541.

Weber, M. (1951). *Gesammelte aufs ä tze zur wissenschaftslehre*. Tübingen.

Wei, J., Wang, X., Schuurmans, D., Bosma, M., Xia, F., Chi, E., Le, Q. V., Zhou, D., et al. (2022). Chain-of-thought prompting elicits reasoning in large language models. *Advances in neural information processing systems, 35*, 24824–24837.

Williams, B. C. (1984). Qualitative analysis of mos circuits. *Artificial Intelligence, 24*(1-3), 281–346.

Williams, M. D. (1978). *The process of retrieval from very long term memory* (tech. rep.). Center for Human Information Processing, UCSD.

Wilson, E. O. (2012). *The social conquest of earth*. WW Norton & Company.

Zelikman, E., Harik, G., Shao, Y., Jayasiri, V., Haber, N., & Goodman, N. D. (2024). Quiet-star: Language models can teach themselves to think before speaking. *arXiv preprint* arXiv:2403.09629.

Zhou, C., Liu, P., Xu, P., Iyer, S., Sun, J., Mao, Y., Ma, X., Efrat, A., Yu, P., Yu, L., et al. (2024). Lima: Less is more for alignment. *Advances in Neural Information Processing Systems, 36*.

Zinn, J. O., & Schulz, M. (2024). Rationalization, enchantment, and subjectivation–lessons for risk communication from a new phenomenology of everyday reasoning. *Journal of Risk Research*, 1–18.